Radiation and the International Space Station

Recommendations to Reduce Risk

Committee on Solar and Space Physics
Committee on Solar-Terrestrial Research
Space Studies Board
Board on Atmospheric Sciences and Climate
Commission on Physical Sciences, Mathematics, and Applications
Commission on Geosciences, Environment, and Resources
National Research Council

NATIONAL ACADEMY PRESS
Washington, D.C.

NOTICE: The project that is the subject of this report was approved by the Governing Board of the National Research Council, whose members are drawn from the councils of the National Academy of Sciences, the National Academy of Engineering, and the Institute of Medicine. The members of the committees responsible for the report were chosen for their special competences and with regard for appropriate balance.

Support for this project was provided by Contract NASW 96013 between the National Academy of Sciences and the National Aeronautics and Space Administration. Any opinions, findings, conclusions, or recommendations expressed in this material are those of the authors and do not necessarily reflect the views of the sponsor.

Cover: Ground track of ISS orbits superposed on a globe along with polar cap areas (shown in yellow), where multimega-electron volt solar energetic particles penetrated to low altitudes during an SPE in November 1997. (Image courtesy of R.A. Leske, R.A. Mewaldt, E.C. Stone, and T.T. von Rosenvinge, "Geomagnetic Cutoff Variations During Solar Energetic Particle Events—Implications for the Space Station," *Proceedings of the 25th International Cosmic Ray Conference*, 2, Space Research Unit, Department of Physics, Potchefstroom University for Christian Higher Education, South Africa, 1997, p. 381.)

International Standard Book Number 0-309-06885-1

Copies of this report are available free of charge from:

Space Studies Board
National Research Council
2101 Constitution Avenue, NW
Washington, DC 20418

Copyright 2000 by the National Academy of Sciences. All rights reserved.

Printed in the United States of America

THE NATIONAL ACADEMIES

National Academy of Sciences
National Academy of Engineering
Institute of Medicine
National Research Council

The **National Academy of Sciences** is a private, nonprofit, self-perpetuating society of distinguished scholars engaged in scientific and engineering research, dedicated to the furtherance of science and technology and to their use for the general welfare. Upon the authority of the charter granted to it by the Congress in 1863, the Academy has a mandate that requires it to advise the federal government on scientific and technical matters. Dr. Bruce M. Alberts is president of the National Academy of Sciences.

The **National Academy of Engineering** was established in 1964, under the charter of the National Academy of Sciences, as a parallel organization of outstanding engineers. It is autonomous in its administration and in the selection of its members, sharing with the National Academy of Sciences the responsibility for advising the federal government. The National Academy of Engineering also sponsors engineering programs aimed at meeting national needs, encourages education and research, and recognizes the superior achievements of engineers. Dr. William A. Wulf is president of the National Academy of Engineering.

The **Institute of Medicine** was established in 1970 by the National Academy of Sciences to secure the services of eminent members of appropriate professions in the examination of policy matters pertaining to the health of the public. The Institute acts under the responsibility given to the National Academy of Sciences by its congressional charter to be an adviser to the federal government and, upon its own initiative, to identify issues of medical care, research, and education. Dr. Kenneth I. Shine is president of the Institute of Medicine.

The **National Research Council** was organized by the National Academy of Sciences in 1916 to associate the broad community of science and technology with the Academy's purposes of furthering knowledge and advising the federal government. Functioning in accordance with general policies determined by the Academy, the Council has become the principal operating agency of both the National Academy of Sciences and the National Academy of Engineering in providing services to the government, the public, and the scientific and engineering communities. The Council is administered jointly by both Academies and the Institute of Medicine. Dr. Bruce M. Alberts and Dr. William A. Wulf are chairman and vice chairman, respectively, of the National Research Council.

COMMITTEE ON SOLAR AND SPACE PHYSICS

GEORGE L. SISCOE, Boston University, *Chair*
CHARLES W. CARLSON, University of California, Berkeley
ROBERT L. CAROVILLANO, Boston College
TAMAS I. GOMBOSI, University of Michigan
RAYMOND A. GREENWALD, Applied Physics Laboratory
JUDITH T. KARPEN, Naval Research Laboratory
GLENN M. MASON, University of Maryland
MARGARET A. SHEA, Air Force Research Laboratory
KEITH T. STRONG, Lockheed Palo Alto Research Center
RICHARD A. WOLF, Rice University

ARTHUR CHARO, Senior Program Officer
RONALD TURNER, Consultant
CARMELA J. CHAMBERLAIN, Senior Project Assistant (through March 1999)
THERESA M. FISHER, Senior Project Assistant (from April 1999)

COMMITTEE ON SOLAR-TERRESTRIAL RESEARCH

MICHAEL C. KELLEY, Cornell University, *Chair*
MAURA E. HAGEN, National Center for Atmospheric Research
MARY K. HUDSON, Dartmouth College
NORMAN F. NESS, Bartol Research Institute
THOMAS F. TASCIONE, Sterling Software

ELBERT (JOE) FRIDAY, JR., Director, Board on Atmospheric Sciences and Climate
TENECIA A. BROWN, Senior Project Assistant

SPACE STUDIES BOARD

CLAUDE R. CANIZARES, Massachusetts Institute of Technology, *Chair*
MARK R. ABBOTT, Oregon State University
FRAN BAGENAL, University of Colorado
DANIEL N. BAKER, University of Colorado
ROBERT E. CLELAND, University of Washington
GERARD W. ELVERUM, JR., TRW Space and Technology Group*
MARILYN L. FOGEL, Carnegie Institution of Washington
BILL GREEN, Former Member, U.S. House of Representatives
JOHN H. HOPPS, JR., Morehouse College
CHRIS J. JOHANNSEN, Purdue University
ANDREW H. KNOLL, Harvard University*
RICHARD G. KRON, University of Chicago
JONATHAN I. LUNINE, University of Arizona
ROBERTA BALSTAD MILLER, CIESIN-Columbia University
GARY J. OLSEN, University of Illinois, Urbana-Champaign
MARY JANE OSBORN, University of Connecticut Health Center
GEORGE A. PAULIKAS, The Aerospace Corporation
JOYCE E. PENNER, University of Michigan
THOMAS A. PRINCE, California Institute of Technology
PEDRO L. RUSTAN, JR., Ellipso, Inc.
GEORGE L. SISCOE, Boston University
EUGENE B. SKOLNIKOFF, Massachusetts Institute of Technology
MITCHELL SOGIN, Marine Biological Laboratory
NORMAN E. THAGARD, Florida State University
ALAN M. TITLE, Lockheed Martin Advanced Technology Center
RAYMOND VISKANTA, Purdue University
PETER VOORHEES, Northwestern University
JOHN A. WOOD, Harvard-Smithsonian Center for Astrophysics

JOSEPH K. ALEXANDER, Director

---

*Former member.

BOARD ON ATMOSPHERIC SCIENCES AND CLIMATE

ERIC J. BARRON, Pennsylvania State University, *Co-chair*
JAMES R. MAHONEY, International Technology Corporation, *Co-chair*
SUSAN K. AVERY, University of Colorado
LANCE F. BOSART, State University of New York at Albany
MARVIN A. GELLER, State University of New York at Stony Brook
DONALD M. HUNTEN, University of Arizona*
JOHN IMBRIE, International Technology Corporation*
CHARLES E. KOLB, Aerodyne Research, Inc.
THOMAS J. LENNON, Brig. Gen. USAF (ret.), WSI Corp.*
ROGER A. PIELKE, JR., National Center for Atmospheric Research
ROBERT T. RYAN, WRC-TV
MARK R. SCHOEBERL, NASA Goddard Space Flight Center
JOANNE SIMPSON, NASA Goddard Space Flight Center
NIEN DAK SZE, Atmospheric and Environmental Research, Inc.
ROBERT WELLER, Woods Hole Oceanographic Institution
ERIC F. WOOD, Princeton University

ELBERT (JOE) FRIDAY, JR., Director

---

*Former member.

COMMISSION ON PHYSICAL SCIENCES, MATHEMATICS, AND APPLICATIONS

PETER M. BANKS, Veridian ERIM International, Inc., *Co-chair*
W. CARL LINEBERGER, University of Colorado, *Co-chair*
WILLIAM F. BALLHAUS, JR., Lockheed Martin Corp.
SHIRLEY CHIANG, University of California at Davis
MARSHALL H. COHEN, California Institute of Technology
RONALD G. DOUGLAS, Texas A&M University
SAMUEL H. FULLER, Analog Devices, Inc.
JERRY P. GOLLUB, Haverford College
MICHAEL F. GOODCHILD, University of California at Santa Barbara
MARTHA P. HAYNES, Cornell University
WESLEY T. HUNTRESS, JR., Carnegie Institution
CAROL M. JANTZEN, Westinghouse Savannah River Company
PAUL G. KAMINSKI, Technovation, Inc.
KENNETH H. KELLER, University of Minnesota
JOHN R. KREICK, Sanders, a Lockheed Martin Company (ret.)
MARSHA I. LESTER, University of Pennsylvania
DUSA M. McDUFF, State University of New York at Stony Brook
JANET L. NORWOOD, U.S. Commissioner of Labor Statistics (ret.)
M. ELISABETH PATÉ-CORNELL, Stanford University
NICHOLAS P. SAMIOS, Brookhaven National Laboratory
ROBERT J. SPINRAD, Xerox PARC (ret.)

NORMAN METZGER, Executive Director (through July 1999)
MYRON F. UMAN, Acting Executive Director (from August 1999)

COMMISSION ON GEOSCIENCES, ENVIRONMENT, AND RESOURCES

GEORGE M. HORNBERGER, University of Virginia, *Chair*
RICHARD A. CONWAY, Union Carbide Corporation (ret.)
THOMAS E. GRAEDEL, Yale University
THOMAS J. GRAFF, Environmental Defense Fund
EUGENIA KALNAY, University of Maryland
DEBRA KNOPMAN, Progressive Policy Institute
KAI N. LEE, Williams College
RICHARD A. MESERVE, Covington & Burling
RADM. JOHN B. MOONEY, JR., USN (ret.), J. Brad Mooney Associates, Ltd.
HUGH C. MORRIS, El Dorado Gold Corporation, Vancouver, British Columbia
H. RONALD PULLIAM, University of Georgia
MILTON RUSSELL, University of Tennessee
THOMAS C. SCHELLING, University of Maryland
ANDREW R. SOLOW, Woods Hole Oceanographic Institution
VICTORIA J. TSCHINKEL, Landers and Parsons
E-AN ZEN, University of Maryland
MARY LOU ZOBACK, U.S. Geological Survey

ROBERT M. HAMILTON, Executive Director

Foreword

A major objective of the International Space Station is learning how to cope with the inherent risks of human spaceflight—how to live and work in space for extended periods. The construction of the station itself provides the first opportunity for doing so.

Prominent among the challenges associated with ISS construction is the large amount of time that astronauts will be spending doing extravehicular activity (EVA), or "space walks." EVAs from the space shuttle have been extraordinarily successful, most notably the on-orbit repair of the Hubble Space Telescope. But the number of hours of EVA for ISS construction exceeds that of the Hubble repair mission by orders of magnitude. Furthermore, the ISS orbit has nearly twice the inclination to Earth's equator as Hubble's orbit, so it spends part of every 90-minute circumnavigation at high latitudes, where Earth's magnetic field is less effective at shielding impinging radiation. This means that astronauts sweeping through these regions will be considerably more vulnerable to dangerous doses of energetic particles from a sudden solar eruption.

This putative radiation danger prompted the present study. It applies what we have learned from past investigations of solar emanations and their effects on Earth's magnetosphere to assess the risk and find ways to minimize it. The study estimates that the likelihood of having a potentially dangerous solar event during an EVA is indeed very high. It also recommends steps that can be taken immediately, and over the next several years, to provide adequate warning so that the astronauts can be directed to take protective cover inside the ISS or shuttle. The near-term actions include programmatic and operational ways to take advantage of the multiagency assets that currently monitor and forecast space weather, and ways to improve the in situ measurements and the predictive power of current models.

The radiation risk is real, but it is also very susceptible to management. That there have been no known overexposures to date is due partly to such good management. Now it is time to revise the protocols and practices of the past to encompass the new challenges of ISS construction and permanent habitation to ensure that this good record continues in the future.

Claude R. Canizares, *Chair*
Space Studies Board

Acknowledgments

Preparing this report would not have been possible without the help of the many individuals who provided the Committee on Solar and Space Physics and the Committee on Solar-Terrestrial Research (CSSP/CSTR) with presentations, consultations, and written materials. The committees are especially grateful to the following individuals: for information on the probability of a solar particle event (SPE) coinciding with an International Space Station (ISS) construction mission, Ronald Turner (ANSER Corporation); for information on the properties of penetrating radiation and on the latitudinal cutoff of SPE particles, Don Smart (Air Force Research Laboratory, ret.); for information on the properties and measurements of highly relativistic electrons in the outer radiation belt, Herbert Kroehl (NOAA's National Geophysics Data Center) and Bernard Blake (Aerospace Corporation); for information pertaining to NOAA's Space Environment Center (SEC), Gary Heckman (NOAA-SEC); for information on matters pertaining to mission operations at Johnson Space Center (JSC) and on radiation data taken on Mir and the shuttles, Michael Golightly and Gautam Badhwar (both at JSC); and for information pertaining to radiation risk management during the Apollo era (the SPAN program), Donald Robbins, Alva Hardy, and Rod Rose (all at JSC during the Apollo era).

In addition, CSSP/CSTR wishes to acknowledge the following individuals for informative presentations and interviews at its meetings: for a flight director's perspective, Paul Hill (JSC); for a flight surgeon's perspective, Jeff Jones (JSC); for an astronaut's perspective, Norman Thagard (Florida State University); for information on radiation effects on biological materials and organisms and on research programs pertaining to these, Walter Schimmerling (NASA headquarters), Richard Setlow (Brookhaven National Laboratory), R.J. Michael Fry (Oak Ridge National Laboratory), and Larry Townsend (University of Tennessee); and for information on radiation effects on other than biological materials and on NASA's Space Environment Effects program at Marshall Space Flight Center, Dana Brewer (NASA headquarters) and Janet Barth (Goddard Space Flight Center). The report has also benefited from inputs by members of the National Research Council's Committee on Space Biology and Medicine, which is chaired by Mary J. Osborn; from the Space Studies Board, especially Fran Bagenal; and from Margaret Kivelson, liaison from the Commission on Physical Sciences, Mathematics and Applications to the Space Studies Board.

This report has been reviewed by individuals chosen for their diverse perspectives and technical expertise, in accordance with procedures approved by the Report Review Committee of the National Research Council (NRC). The purpose of this independent review is to provide candid and critical comments that will assist the authors and the NRC in making the published report as sound as possible and to ensure that the report meets institutional standards for objectivity, evidence, and responsiveness to the study charge. The contents of the review comments and draft manuscript remain confidential to protect the integrity of the deliberative process. CSSP/CSTR wishes to thank the following individuals for their participation in the review of this report: J. Bernard Blake, the Aerospace Corporation; Joan Feynman, Jet Propulsion Laboratory; R.J. Michael Fry, Oak Ridge National Laboratory; John Grunsfeld, NASA Johnson Space Center; Louis J. Lanzerotti, Lucent Technologies; Edward T. Lu, NASA Johnson Space Center; Frank B. McDonald, University of Maryland; and Donald J. Williams, Johns Hopkins University Applied Physics Laboratory.

Although the individuals listed above have provided many constructive comments and suggestions, responsibility for the final content of this report rests solely with CSSP/CSTR and the NRC.

Contents

EXECUTIVE SUMMARY		1
1	SCOPING THE PROBLEM	7
	1.1 Radiation in Space, 7	
	1.2 Space Weather Context, 10	
	1.3 Metrics of Radiation Risk, 11	
	1.4 Radiation and the International Space Station, 12	
	1.5 Issues in Managing Radiation Risk During ISS Construction, 18	
	1.6 The Apollo Experience, 19	
	1.7 Summary and Recommendation, 20	
	1.8 Notes and References, 21	
2	SOLAR PARTICLE EVENTS AND THE INTERNATIONAL SPACE STATION	24
	2.1 Background to an Assessment of SPE Impacts on ISS Construction, 24	
	2.2 Probability of SPE Impact on ISS Construction, 25	
	2.3 Correlation Between SPEs and Size of the SPE Zone, 27	
	2.4 Summary and Recommendation, 30	
	2.5 Notes and References, 30	
3	RELATIVISTIC ELECTRONS AND THE INTERNATIONAL SPACE STATION	32
	3.1 Outer Belt Electrons, 32	
	3.2 Monitoring Outer Belt Electrons, 33	
	3.3 Predictability of Radiation Belt Electrons at Low Altitude, 34	
	3.4 Assessment of Hazards Faced by Astronauts During ISS Construction, 35	
	3.5 Operational Strategy, 36	
	3.6 Summary and Recommendations, 37	
	3.7 Notes and References, 38	

4 SPACECRAFT SOURCES OF OPERATIONAL RADIATION DATA — 39
4.1 Value of Spacecraft Monitors in Support of ISS Construction, 39
4.2 An Interagency Fleet of Spacecraft Monitors, 41
4.3 Future Spacecraft in Support of ISS Operations, 42
4.4 Summary and Recommendation, 44
4.5 Notes and References, 44

5 INTERAGENCY CONNECTIONS — 45
5.1 Institutional Factors Limiting Interagency Ability to Provide Better Information for Operational Radiation Risk Assessments, 45
5.2 Recommendations, 47

6 INTRA-NASA CONNECTIONS — 48
6.1 Radiation: A Concern Throughout NASA, 48
6.2 NASA Programs That Involve Radiation, 49
6.3 Communication Between Programs with an Interest in Radiation, 50
6.4 Summary and Recommendation, 51

EPILOGUE: A NOTIONAL SCENARIO FOR IMPROVED SUPPORT OF INTERNATIONAL SPACE STATION CONSTRUCTION — 52
E.1 Vision of an ISS Construction Mission Supported by Reliable, Accurate Radiation Forecast Models During the Solar Maximum, 52
E.2 The Way Things Ought to Work, 52
E.3 The Missing Pieces, 55
E.4 Timetable for Implementing the Report's Recommendations, 55

APPENDIXES

A SPACE WEATHER MODELS APPLIED TO RADIATION RISK REDUCTION — 59
A.1 Space Weather Models, 59
A.2 Near-Earth Space Environment Models, 61
A.3 Advanced Empirically Based Forecast Models of Radiation Risk Parameters, 62
A.4 Observational Supplements to Model-Based Forecasts, 62
A.5 National Space Weather Program, 63
A.6 Summary and Findings, 64
A.7 Notes and References, 66

B STATEMENT OF TASK — 68

C BIOGRAPHIES OF COMMITTEE MEMBERS — 70

D ACRONYMS AND ABBREVIATIONS — 74

Executive Summary

INTRODUCTION

This report originated with a request from the National Aeronautics and Space Administration (NASA) (Appendix B). To construct the International Space Station (ISS) and maintain it during construction, astronauts and cosmonauts will work in space suits outside their spacecraft in shifts, each of which is projected to last for 6 hours, for a total amount of time estimated to exceed 1,500 hours. According to the present construction schedule, these extravehicular activities (EVAs) will occur over a 4-year period that straddles the peak in activity of the current solar cycle. After the 4-year period, one or two EVAs per month will probably continue for the life of ISS. The peak in the solar cycle combines with the station's high-inclination orbit to add a new concern for managers of radiation risk.

Unlike the originally planned low-inclination orbit (28 degrees), the finally approved high-inclination orbit (51.6 degrees) cuts through high-latitude radiation environments that are sometimes quite harsh, as was noted when the redesign was contemplated in the early 1990s. These high-latitude radiation environments (energetic particles from solar storms and relativistic electrons in Earth's outer radiation belt) vary greatly over time, from benignly calm to severely stormy. At the height of their storminess, they can be intense enough to pose a radiation hazard to astronauts engaged in EVAs, although doses estimated for even worst-case scenarios fall short of life-threatening. The principal risk to astronauts that increased exposure to radiation in ISS orbit poses is the increased probability of developing cancer later in life. The principal concern for flight directors that increased exposure of astronauts to radiation raises is the potential impact on flight schedules and crew rotation if a radiation event pushes an astronaut over an allowable radiation limit. Astronauts are also concerned that crossing an allowable radiation limit will restrict flight opportunities. Storms bearing intense radiation are relatively rare, but EVAs during ISS construction flights are relatively frequent, which raises a concern that the two might by chance coincide. Information obtained during the course of this study puts at near-certainty the likelihood that on one or more occasions an ISS construction flight will be in progress when a high-latitude radiation event (described below) occurs.

This finding naturally raises the question, What is the status of radiation risk management as it pertains to ISS construction? It would seem to be a simple matter, for example, for the Space Environment Center (SEC) of the National Oceanic and Atmospheric Administration (NOAA) or for NASA's own satellites to identify solar events that could cause radiation problems and to get such information to the flight director in time to take appropriate action. But an overly restrictive flight rule and the lack of operationally calibrated models bar the path between the flight director and such sources of information. The problematic (albeit unofficial) flight rule is the "real-time, on-site data" rule, which says that changes in flight plans in response to a radiation situation must be based on real-time, on-site data only. The first recommendation of CSSP/CSTR addresses this flight rule.

Recommendation 1: Because it denies access to valid information and thus unnecessarily restrains flight-director options, flight directors should not adhere rigidly to the (unofficial) real-time, on-site data rule.

As mentioned, the second obstacle in the path between the flight director and data sources is the lack of operationally calibrated models. In important cases, however, the state of radiation modeling is advanced enough, or with directed effort could quickly become advanced enough, to justify a flight rule that allows use of validated procedures to infer and, in some cases, to predict on-site radiation conditions from off-site data. The report cites such cases.

CSSP/CSTR notes that Russians performing EVAs will be directed out of the Russian mission control center in Moscow. Further, it is likely that U.S. and international crew members on ISS will also participate in EVAs directed out of mission control-Moscow. However, flight rules at mission control-Moscow pertaining to radiation may differ from those at NASA's mission control center. Although this report is directed at NASA, CSSP/CSTR believes that some of its recommendations could also be implemented by mission control-Moscow.

SOLAR PARTICLE EVENTS AND THE INTERNATIONAL SPACE STATION

Based on the assumption—the best now available—that the radiation characteristics of the current solar cycle will resemble those of the last cycle, there is nearly a 100 percent chance that at least 2 out of 43 planned ISS construction flights will overlap a significant solar particle event (SPE) and a 50 percent chance that at least 5 flights will overlap such an event. Moreover, the high-latitude zones to which solar energetic particles have access show a marked tendency to widen over the polar latitudes reached by the ISS orbit when SPEs are in progress, a tendency that becomes more pronounced as SPEs intensify. Two storms during 1989, near the maximum of the last solar cycle, illustrate the point. The areas around the poles accessible to SPE particles enlarged until they engulfed more than a quarter of the ISS orbit, and the flux of particles was high enough to have pushed an astronaut over the short-term limit for irradiation of skin and eyes during a single ill-timed 6-hour EVA. These results would seem to call for an aggressive program aimed at reducing solar radiation risk to astronauts during ISS construction. Recommendation 2 addresses means of implementing the elements of such a program.

Recommendation 2: For real-time SPE risk management, carry out the steps needed to make usable by SEC and the Space Radiation Analysis Group (SRAG) at Johnson Space Center (JSC) models that use real-time data to specify the intensity of SPE particles and the geographical size and shape of the zones accessible to them.

NASA, NOAA, the U.S. Air Force (USAF), and the distributed space physics community have the capability for implementing this recommendation. The project implied in this recommendation is one of the important projects that could be implemented early enough to have an impact on SPE radiation risk management during ISS construction. It should receive high priority for early implementation. (Appendix A discusses a suite of models for this application.)

RELATIVISTIC ELECTRONS AND THE
INTERNATIONAL SPACE STATION

For a portion of nearly every day, some fraction of the ISS orbit lies within the outer radiation belt, where relativistic electrons reside. At its maximum, this fraction is about 20 percent. During occasions called relativistic electron events, which happen on average about once per month and last several days, the intensity of relativistic electrons in the belt increases by up to four orders of magnitude. When the intensity of relativistic electrons is greatest, a single ill-timed EVA could deliver a radiation dose big enough to push an astronaut over the short-term limit for skin and eyes. To minimize the possibility of scheduling EVAs during such events, procedures can be implemented to specify and forecast at least approximately the intensity of relativistic electrons in the outer belt. NOAA Polar-Orbiting Operational Environmental Satellites (POES) provide measurements of relativistic electron fluxes that can be used to calculate with reasonable accuracy the relativistic electron environment at ISS. These measurements are available only about every hour and a half, however. NOAA Geostationary Operational Environmental Satellites (GOES), on the other hand, provide relativistic electron measurements continuously, but these measurements are not so directly transferable to the ISS orbit. Nonetheless, the intensity of GOES measurements tracks the intensity of POES measurements in the outer belt. Thus, in combination, POES and GOES measurements would allow radiation risk managers to quantitatively follow variations of electron intensity in the outer belt. A crucial piece of hardware that the ISS project should provide is an electron dosimeter attached outside the station. This would allow SRAG to test the quality of the specifications and forecasts that are possible from measurements taken by POES and GOES. These considerations lead to three related recommendations.

Recommendation 3a: NASA should implement a procedure for using POES and GOES measurements of relativistic electrons in the outer radiation belt to specify and forecast the electron radiation environment at ISS. (Such a procedure is outlined in Section 3.3.)

Recommendation 3b: As soon as possible, JSC should install an electron dosimeter and an ion dosimeter outside the ISS that can return data in real time to SRAG at JSC.

Recommendation 3c: A project should be initiated to develop a protocol for identifying the conditions that produce highly relativistic electron events based on the demonstrated good correlation between changes in solar wind conditions and the onset of such events. The recommended project might be a candidate for support by one of the affiliated agencies of the National Space Weather Program (NSWP). (See Section A.5.)

SPACECRAFT SOURCES OF OPERATIONAL RADIATION DATA

Data that could contribute to reducing radiation risk are currently being acquired by a strategically placed multiagency fleet of research and operational spacecraft. This fleet can provide information on the radiation environment of ISS orbit in real time and in advance of real time. Some spacecraft monitor the Sun and its corona in multiple wavelengths and so can diagnose the X-ray potency of solar flares and warn of oncoming material spewed from the Sun by solar storms. They also monitor SPE fluxes to give direct information on the radiation intensity within zones accessible to SPE particles. Other spacecraft monitor solar wind parameters, which can be used to compute the size and shape of SPE-accessible zones. Spacecraft in relatively low-altitude, polar orbits monitor the flux of relativistic electrons in the outer radiation belt, which the ISS orbit transects. Recommendation 4 addresses a need to channel the relevant information to SRAG at JSC.

Recommendation 4: Promptly convene a meeting of pertinent NASA Space Science Enterprise, SRAG, and SEC managers with the principal investigators of satellite instruments. The meeting would

(1) consider ways to extend the capabilities of the current spacecraft fleet to provide real-time radiation data for driving models and specifying the ISS radiation environment and (2) formulate an implementation plan for swiftly channeling such data to radiation risk managers at JSC.

INTERAGENCY CONNECTIONS

A major obstacle stands in the way of implementing any of the improved scientific data and modeling resources that are currently available from research programs in NASA and the National Science Foundation (NSF). Both SEC and SRAG are fully tasked in maintaining their daily program of data collection and analysis. Ongoing operations require that these be maintained. Incorporation of improvements thus becomes a secondary activity, and the lack of adequate resources and agency support in both organizations limits the rate at which improvements can be made. The next two recommendations address this condition.

Recommendation 5a: NASA, NOAA, and the USAF should cooperate to support the activities that would lead to an operational space weather forecasting capability.

Recommendation 5b: NOAA should extend the range of its SPE predictions from the present ≥10 MeV to biologically effective energy ranges. Forecasts of particle energies at several steps between 10 and 100 MeV would be a significant improvement for space radiation use as well as for other users who operate satellites and systems in space.

INTRA-NASA CONNECTIONS

There are other major programs at NASA besides the manned flight program that require an accurate knowledge of Earth's radiation environment. The kind of knowledge required varies from program to program, but the range of knowledge needed extends from the basic science, physical processes, and generation mechanisms of the radiation belts and particle events, to net integrated radiation doses averaged over a long period of time. The trend in recent years at NASA has been toward smaller and cheaper spacecraft with heavy use of microelectronics, smaller instruments, and more onboard data processing. For this and other reasons, the working knowledge of Earth's radiation environment (models, forecasts of particle events and disturbances, integrated doses, etc.) needs to be improved to address current planning and development requirements in just about every area of NASA activity.

Recommendation 6: To coordinate intra-NASA activities and concerns related to radiation, NASA should establish an agency-level radiation plan and task force. It should also establish a multidisciplinary steering committee to advise the task force.

SPACE WEATHER MODELS APPLIED TO RADIATION RISK REDUCTION

The above recommendations address means to exploit currently available resources to allow a rapid augmentation of the tools available for radiation risk management during ISS construction. Looking beyond these recommendations, there are actions of a tactical nature that can be taken to strengthen radiation risk management in the late phases of ISS construction and during ISS operations. These actions entail the selective implementation of space weather modeling efforts. Space weather modeling is the discipline responsible for developing models that take information from where instruments happen to be and use it to specify and forecast the space environment at places where the information is wanted.

Appendix A identifies research projects that address specific elements of an effort that would provide high-quality information on the parameters most crucial to assessing radiation risk. Two projects deserve particular attention, the first because its potential impact on radiation risk reduction is especially crucial, the second because the maturity of its models promises early, tangible results:

1. *Mapping latitudes at which SPE particles can penetrate under a variety of geomagnetic conditions to the altitude of ISS.* Several methods have been proposed; these should be critically evaluated and the best should be implemented. (Appendix A describes a suite of such methods.)
2. *Developing operational space weather nowcast and forecast codes.* Several of the existing advanced, data-based space weather nowcast and forecast codes (see Appendix A) could be transformed relatively quickly into operational codes to give SRAG the ability to forecast at least some radiation-risk parameters during most of the ISS construction period.

These projects could be undertaken in the near term by one or more of the affiliated agencies of the NSWP (see Section A.5 for a description of NSWP).

TIMELINE FOR IMPLEMENTING RECOMMENDATIONS

Figure ES.1 gives a timeline keyed to the ISS construction schedule (as it was known in July 1999) for the implementation of the recommendations of this report. As shown in the figure, there are recommendations that should be implemented immediately (R1, R3a, R4, R5b, and R6), recommendations that will require 1 or 2 years

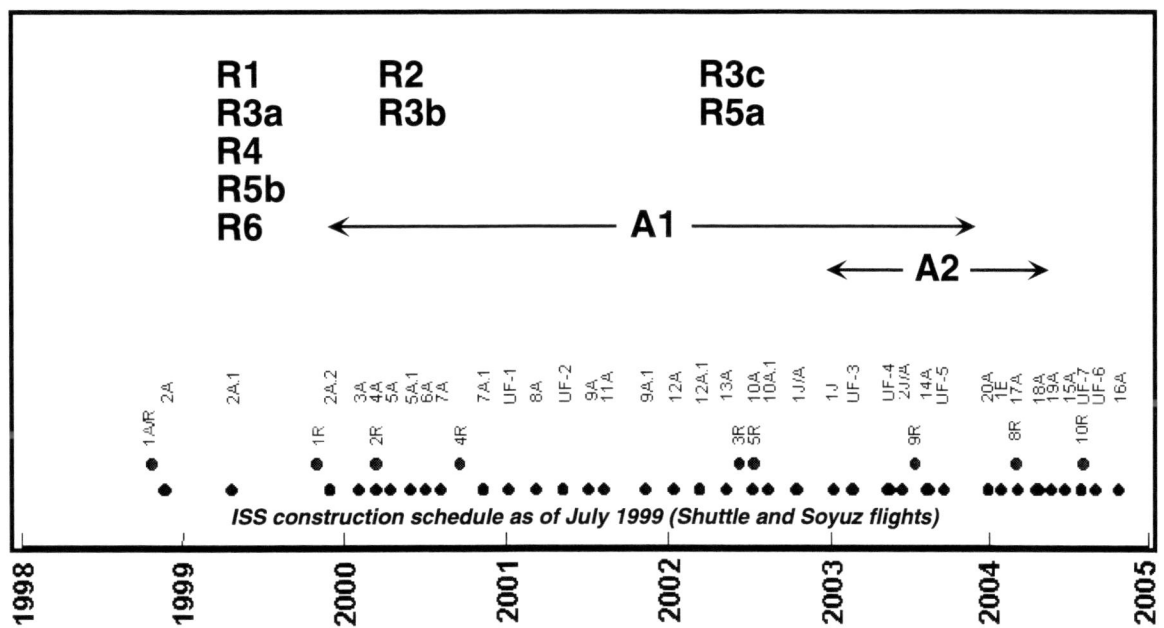

Figure ES.1 Timeline for implementing the report's recommendations, which are denoted R1 through R6. A1 and A2 refer to priority research activities (see Appendix A). The timeline is keyed to the ISS assembly sequence, which is available from NASA on the Web at <http://spaceflight.nasa.gov/station/assembly/flights/chron.html>. The first crew—a U.S. astronaut and two Russian cosmonauts—will be launched on a Russian Soyuz spacecraft in March 2000 on flight 2R (the letter "R" in the flight designation denotes a Russian mission). They will stay there 3 months. From that point on, the ISS is planned to be permanently staffed.

to implement (R2 and R3b), and recommendations that will take several years to implement (R3c and R5a). The research needed to improve space weather services in support of manned missions (described in Appendix A) is also shown in Figure ES.1. Research activities are organized into two groups: (1) those that can be implemented within 5 years (A1), with some of these activities being implemented within 1 year, and (2) those requiring more time to implement (A2).

1

Scoping the Problem

1.1 RADIATION IN SPACE

"My God, space is radioactive!" Recoiling thus from the 1958 discovery of Earth's radiation belts, Ernie Ray, a Van Allen protégé, gave generations of space workers a motto and a challenge. For scientists the challenge is to map the radiation of space, to document its behavior, and to understand its causes. For engineers it is to design radiation-tolerant space systems and hardware. For managers it is to deploy scientific results and engineering capability to build space programs that survive despite radiation.

The problem is serious. Over the past 20 years, radiation effects have caused between one and two satellites per year on average to suffer total or partial mission loss.[1] Satellites at low latitudes in low Earth orbit (LEO) stay relatively safe by ducking the intense heart of the radiation belts higher up. But at higher altitudes and higher latitudes, where Earth's radiation belts reside and radiation from solar storms invades, radiation hazard cannot be ignored. James Michener, in his book *Space*, describes the fictional death of two astronauts on the Moon from radiation emitted by a solar storm. The scenario is accurate regarding the radiation hazard in high-altitude space, although it exaggerates doses. (Units of radiation dose are described in Section 1.3.) The scene might have been inspired by the great solar storm of August 4, 1972, which like William Tell's apple-splitting arrow, split the 8 months between the last two Apollo Moon landings evenly. It delivered a total dose of radiation over half a day that, had it missed the middle and hit the Apollo mission at either end, would have caused the crew in the lunar module to suffer acute radiation sickness and, given the uncertainty in the estimate, possibly even death.[2]

The International Space Station (ISS)[3] will be exposed to penetrating particle radiation from three sources: terrestrial, solar, and galactic. The terrestrial source is Earth's radiation belts (a.k.a. the Van Allen belts), of which four are recognized: the inner and outer ion belts and the inner and outer electron belts. Because they reach energies that penetrate matter to significant depths, ions in the inner belt and electrons in the outer belt pose the greatest hazards to space hardware and astronauts. Figure 1.1 shows where in general the belts occur in space. Shaped in cross section like a crescent concave earthward, they peak in altitude over the equator and project down like horns to high latitudes. In this they conform to the geomagnetic field, whose lines of force Figure 1.1 depicts. The "equator" over which the belts peak is the magnetic equator, which tilts from the geographic equator by 11.4 degrees, deviating most over Peru and Sumatra. Moreover, the lines of magnetic force are not geocentric but instead are offset such that the belts come closest to Earth over the South Atlantic, giving rise to a feature known as the South Atlantic Anomaly (SAA). Consequently, whereas low-latitude satellites in low-altitude Earth orbit

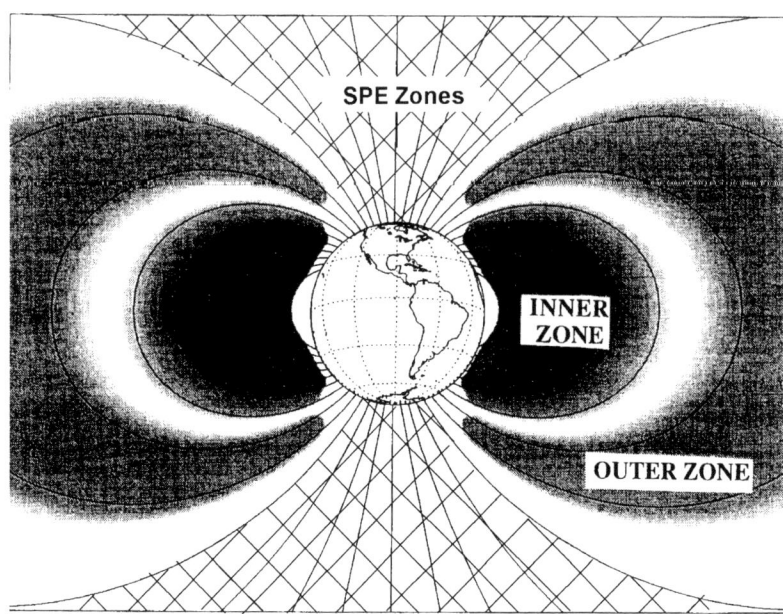

Figure 1.1 The radiation environments of the International Space Station (adapted from Baker[4]). The figure shows the three regions of space around Earth where penetrating radiation occurs. The inner and outer radiation belts each have an electron and ion component.

(LEO satellites) can fly under the inner belt in most places, they cannot avoid it within the SAA. (Note: except for those in equatorial orbits, which largely avoid the SAA.) Based on data from the Mir space station, cosmonauts and astronauts normally accumulate about half of their total radiation dose during the 2 to 5 percent of the time they spend in the SAA. High-inclination LEO satellites also encounter the polar caps, which are accessible to solar energetic particles. Figure 1.2 shows locations of radiation-induced upsets ("hits") suffered by computer memory in a polar-orbiting LEO satellite (UOSAT-2). The hits generate a pointillistic map of the SAA and the horns of the belts at LEO altitude. A final point to note regarding the belts, especially the outer belt, is that their intensity varies with time. The outer electron belt shows variations synchronized with distinctive and relatively common solar wind conditions known as high-speed solar wind streams. During such conditions, the intensity of energetic electrons can increase by many orders of magnitude. Space physicists call times of elevated intensities of energetic electrons highly relativistic electron events (HRE events).

Penetrating particle radiation from the Sun takes the form of solar particle events (SPEs), which typically last several days to a week. Because penetrating SPEs are mainly composed of protons generated by solar storms, they share the statistical properties of these storms. They exhibit a quasi-11-year cycle loosely synchronized with the solar activity cycle. During the last solar cycle (cycle 22), 20 SPEs were officially designated as such by NOAA's Space Environment Center (SEC). (Section 2.1 includes the criterion SEC uses to declare an SPE to be in progress.) The SEC, one of NOAA's eight national centers for environmental prediction (NCEPs), is responsible for the space environment. SPEs constitute the high-altitude acute radiation hazard. The geomagnetic field shields low-latitude LEO satellites from SPEs. Shielding ceases, however, at altitudes above about 4 Earth radii (1 Earth radius, or Re = 6,370 km) or at LEO latitudes (geomagnetic) above about 60 degrees. Geomagnetic storms, which are terrestrial responses to solar storms, weaken this shielding and allow solar energetic particles to penetrate to lower altitudes and latitudes. Figure 1.1 portrays the situation during a storm, in which solar energetic particles fill the space above 50 degrees geomagnetic latitude. In rough numbers, one large SPE can impart to a high-

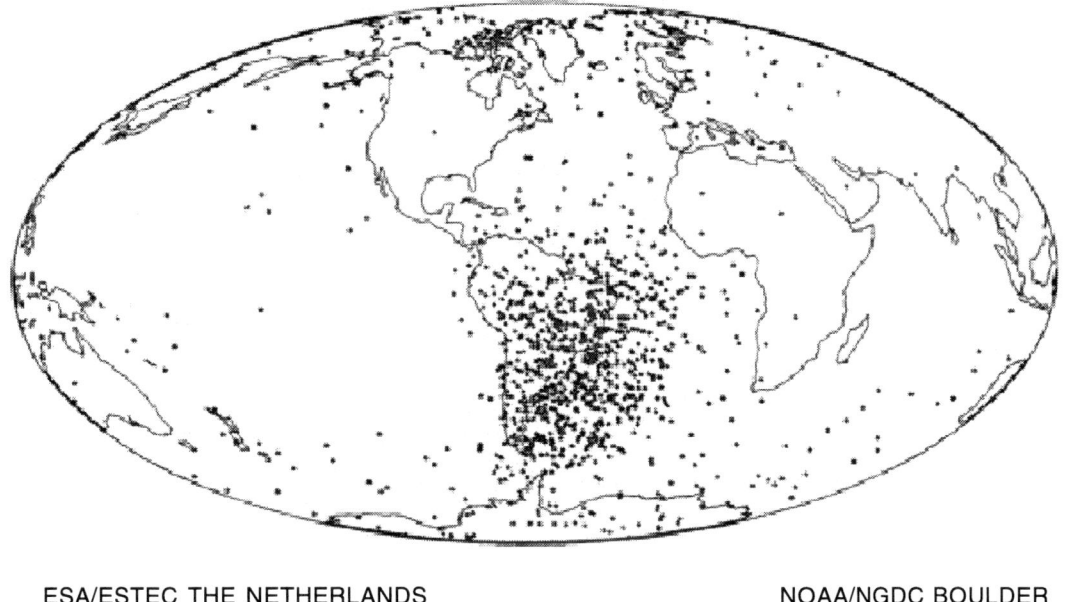

ESA/ESTEC THE NETHERLANDS   NOAA/NGDC BOULDER

Figure 1.2 Locations of single-event memory upsets suffered by the UOSAT-2 outlining the regions of the South Atlantic Anomaly and the outer radiation belt (from CSSP/CSTR[5]). (Courtesy M.A. Shea.)

inclination LEO satellite a dose comparable to the SAA dose accumulated by the satellite over about 100 days. If such a dose were absorbed by an astronaut in a space suit, it would be equivalent to about 1 year's accumulated SAA dose inside a space cabin. This generalization is quantified in Section 1.4.

Galactic cosmic rays (GCRs) make up the third form of radiation in space. They present a low-level, continuous source of penetrating radiation. They are partially shielded by the geomagnetic field. Low-inclination LEO satellites receive about half the GCR dose received by high-inclination LEO satellites. The accumulated GCR dose to LEO satellites is, as a rule, less than or comparable to the accumulated SAA dose and is rather insensitive to the level of solar activity.

Figure 1.3 shows how the penetrating power of energetic protons, the main constituent of SPEs, increases with increasing proton energy. It marks the thicknesses of the different parts of a space suit at mid-deck within a space shuttle and within the Mir space station (R-16 dosimeter and Lulin dosimeter). Protons of only 10 MeV energy can penetrate nearly three-quarters of the surface area of a space suit. It takes a 25-MeV proton to penetrate the most heavily shielded part, the visor. Above 30 MeV, protons can penetrate the mid-deck of the space shuttle. Arrows show the energies measured by two dosimeters inside Mir. The 10 MeV threshold for penetrating a space suit is also the energy that forecasters at SEC monitor to watch for the onset of an SPE. A similar energy curve for electrons shows that 0.5 MeV particles, an energy characteristic of HRE events, can penetrate space suits. HRE events typically have high flux levels, between half a million and several million electron volts, that penetrate space suits, although they rarely have dangerous fluxes of electrons at energies that can penetrate a shuttle or station hull. **This report focuses on solar energetic particles with energies higher than 10 MeV and outer radiation belt electrons with energies higher than 0.5 MeV. These are the energies at which protons and electrons penetrate space suits. Because they are transient and hard to predict, these populations of penetrating particles pose the greatest challenge to ISS radiation risk managers.** By comparison with highly variable SPEs and HRE events, the relatively stable SAA is well understood and predictable.[6] This report does not therefore address the SAA in any detail.

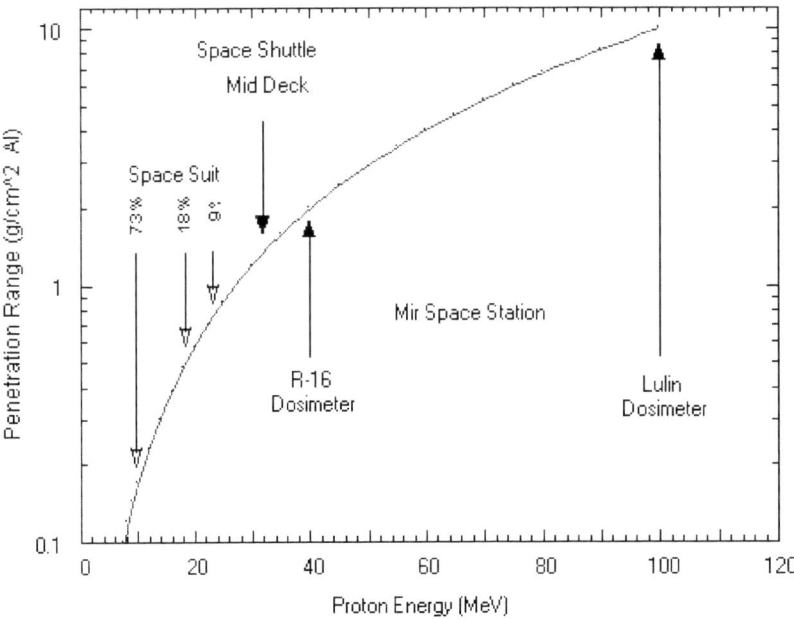

Figure 1.3 Range-energy curve for protons showing energies capable of penetrating various items of space hardware. (Courtesy of Don Smart.)

1.2 SPACE WEATHER CONTEXT

This section gives a short overview of the phenomena that make up space weather.[7] The term "space weather" refers to conditions on the Sun and in the solar wind, magnetosphere, ionosphere, and thermosphere that can influence the performance and reliability of spaceborne and ground-based technological systems and that can affect human life or health.[8] The Sun ultimately drives all space weather phenomena by means of the solar wind (the hot, magnetized plasma flowing nearly radially outward from the Sun at all times) and solar storms (flares, prominence eruptions, and coronal mass ejections (CMEs)). Long-term trends in space weather, which are related to the 11-year solar activity cycle, are far easier to predict than short-term variations such as individual solar storms. This is analogous to our ability to forecast Earth's atmospheric weather: we know that North America will be relatively cold next winter and warm next summer but find it difficult to predict next week's weather. Similarly, solar storms are more frequent during the years around the solar cycle maximum than during the solar minimum. Because the next solar maximum is upon us (activity is expected to peak in 2000-2001), periods of potentially dangerous space weather are likely to be more frequent in the next few years. This increased space-weather hazard coincides with the peak construction period for ISS.

Until a decade ago, the primary driver of adverse space weather was thought to be solar flares. These explosive outbursts had been observed from Earth, first in optical wavelengths and then in radio and X-ray emissions, for over a century, as had been the associated ionospheric and geomagnetic disturbances. They were thus the only likely candidates until the discovery of CMEs by the first spaceborne coronagraphs in the 1970s.[9-11] These huge eruptions appeared at first to be powered by flares. However the ensuing wealth of solar and interplanetary data collected by spaceborne instruments, of higher resolution and sensitivity than ever before, led to a fundamental realization: flares play a role secondary to CMEs in the initiation and propagation of geoeffective disturbances.[12, 13] The exact relationship between flares and CMEs is complex and the subject of much debate at present; observations indicate that some flares may be initiated by CMEs, but the reverse is rarely if ever true.

As the eruption traverses the heliosphere, the magnetic topology of the CME (and the associated shock that is typically present for a "fast" CME) continue to evolve owing to variations in the surrounding plasma and magnetic field characteristics, compression and draping of the interplanetary field around the propagating disturbance, and interactions with preceding (slower) or following (faster) solar storms and solar-wind structures. Consequently, the configuration of the magnetic disruption that reaches Earth can be much different from its initial configuration near the Sun. Thereafter, the essential factor in determining the severity of a geomagnetic storm is the degree of magnetic misalignment at the point of impact between the solar eruption and Earth's self-generated magnetic field, called the magnetosphere (see below).

Of prime relevance to the core issue of this report is the discovery of the role played by shocks that are caused by fast CMEs. The most energetic SPE particles can reach Earth within 10 to 100 minutes of the solar manifestations of the storm seen in visible or X-ray wavelengths. The time delay depends on whether the source of the particles is a flare or a shock and on the heliographic location of the source as seen from Earth, so particle arrival times vary widely. Moreover, their flux and energy spectra evolve in transit owing to continued acceleration and transport effects.

Earth's magnetosphere is confined and shaped by the magnetized solar wind. As the solar wind passes Earth, it severely compresses the magnetospheric field on the dayside and draws it out into a long, cometlike tail (the magnetotail) on the nightside. Many of the field lines that thread the magnetotail are "open": that is, they connect Earth's polar-cap ionosphere to the interplanetary medium and thus ultimately to the Sun or to interstellar space. In contrast, closed magnetospheric field lines have both "feet" on Earth, one in the southern hemisphere and one in the northern, and do not extend into the interplanetary medium.

SPE particles can reach low Earth altitudes directly by spiraling along open field lines. Solar-particle access is not limited entirely to the open field lines of the polar cap, however, for very energetic solar protons can leak onto closed field lines near the polar-cap boundary between open and closed magnetic flux. The more energetic the proton, the farther it can penetrate onto closed field lines. During geomagnetic storms, the solar wind or CME compresses the magnetosphere more severely, more field lines open as a result of magnetic reconnection, and the polar cap grows. Solar particles gain access to larger regions above Earth, particularly at the highest magnetic latitudes, which will be traversed by the ISS. Thus, the dynamic response of the magnetosphere to solar disturbances will increase the solar particle fluxes encountered by ISS.

Earth's radiation belts are composed of trapped energetic particles, which pose an additional radiation hazard for ISS. One component of the belts, the outer-belt MeV electrons, has long been known to be highly variable. Fluxes frequently vary by several orders of magnitude, with an interval of high flux observed typically once a month. Though MeV electrons rarely penetrate into the interior of a spacecraft, they can be hazardous to astronauts performing EVAs. Most of the other components of the belts, including highly penetrating energetic protons, are generally much more stable and predictable. However, we now know that that stability is not absolute. Radiation-belt experts were taken by surprise when on March 24, 1991, the Combined Release and Radiation Effects (CRRES) spacecraft observed a substantial energization and reorganization of the belt structure, including the energetic protons, when an exceptionally strong interplanetary shock hit Earth. A few qualitatively similar but weaker events have subsequently been found by retrospective analysis of old data. Thus, both major components of the radiation hazard faced by ISS—solar and radiation-belt particles—are strongly affected by the overall dynamics of the magnetosphere and particularly by the way in which the magnetosphere reacts to extreme events in the heliosphere.

1.3 METRICS OF RADIATION RISK

How one measures radiation depends on the application. For biological applications, the quantity of interest is the radiation dose absorbed by living tissue, for which the standard units are the gray (1 Gy = 1 joule of radiation energy absorbed per kilogram of tissue) and the centigray (0.01 Gy), also called a rad. For a given dose in these units, the biological effects vary with the type of radiation. A dose of energetic particles normally causes more

Table 1.1 NCRP Recommended Dose Limits for All Organs and Ages (in sieverts)

Limit	Bone Marrow	Eye	Skin
Thirty-day	0.25	1.0	1.5
Annual	0.5	2.0	3.0
Career	See Table 1.2	4.0	6.0

Table 1.2 Updated 1999 NCRP Recommended Career Dose Limits Based on 3 Percent Lifetime Risk of Induced Cancer (in sieverts)

Age at Exposure	Female	Male
25	0.5	0.8
35	0.9	1.4
45	1.3	2.0
55	1.7	3.0

damage than the same dose of energetic photons (X-rays or gamma rays). Particles with high atomic numbers and high energy (HZE particles) cause the greatest damage for a given dose. Units designed to measure the relative biological effectiveness (RBE) of radiation are the sievert (Sv) and the centisievert (1 cSv = 0.01 Sv), also called a rem. These are obtained from the gray and rad, respectively, by multiplying by an experimentally determined quality factor. The quality factor is defined to be unity for gamma rays. Thus, it measures the excess or deficit of radiation damage as a proportion of the gamma-ray damage for the same dose. Quality factors for the SAA vary from 1.6 to 1.9, depending on the shielding, and for GCRs, they vary from 2.9 to 3.5, higher values being associated with higher inclination orbits.[14] These values apply inside a shuttle or space station.

The biological effects of radiation may be classified as acute radiation sickness, late deterministic effects, and stochastic effects. Acute radiation sickness follows exposure to a radiation dose generally greater than 1 Sv in a time generally less than 1 day. Depending on dose, symptoms, which include nausea and vomiting, start within a few hours to a day.[15] Symptomatic reactions to acute radiation doses vary greatly between individuals, so that dose effects must be expressed probabilistically. The probability of vomiting within 2 days is about 10 percent for an abdominal dose of 1 Sv. It reaches about 90 percent for a dose of 3 Sv.[16] The probability of death is about 10 percent for a whole-body 3-Sv dose and reaches about 90 percent for a whole-body 4-Sv dose.[17] Whole-body dose means a dose to the blood-forming organs (mainly bone marrow). These probabilities refer to situations in which no countermeasures are taken. Countermeasures are quite effective at ameliorating symptoms of radiation sickness.

Late deterministic effects from radiation include cataract formation and temporary sterility. There are radiation thresholds for damage to tissues in various critical organs such as bone marrow, lenses of the eye, and skin. These threshholds form a basis for establishing guidelines for short-term (30-day and yearly) radiation dose limits. The risk for induced cancer appears to have no threshold below which it vanishes; it simply decreases as the accumulated dose decreases. Recommended career limits on accumulated radiation dose are therefore set by considerations of an acceptable increase in the risk of cancer. In a 1989 report, the National Council on Radiation Protection and Measurements (NCRP) published recommended limits on short-term and career limits.[18] The career limit was based on a 3 percent risk of induced cancer. A recent update of this report suggests reducing the recommended career limit, still based on a 3 percent risk of induced cancer, by a factor of two. The new recommended career limits are in line with international limits set for workers in terrestrial radiation environments. Other limits in the NCRP report were judged still valid. It is believed that if these limits are observed, no acute or late deterministic effects will develop. Tables 1.1 and 1.2 give the current (updated in 1999) NCRP recommended dose limits. These are low-dose-rate values, which are appropriate to SPE situations.

1.4 RADIATION AND THE INTERNATIONAL SPACE STATION

As originally conceived in the early 1980s, ISS was to have a low-inclination (28.5 degrees), low-altitude (350 km) orbit. Then, the SAA and GCRs would have been the only significant sources of radiation. SRAG knows how to design mission schedules to minimize astronaut exposure to the SAA during EVAs, and there is little anyone can do to minimize GCR exposure. In 1993, however, the United States agreed with the Russian Federation to incorporate Russian launch capabilities into ISS construction and maintenance. That agreement brought with it the need to place ISS in a high-inclination orbit, essentially the same as that of the Mir space station, 51.6 degrees geographic. Consequently, ISS and the astronauts who construct and use it run the risk of being exposed to solar energetic particles and penetrating electrons in the horns of the outer belt. Exposure from these sources will be sporadic since SPEs follow solar storms and HREs follow magnetic storms and impacts by strong solar wind shocks. During the declining phase of a solar cycle—perhaps late in ISS construction—HRE events are also associated with times, lasting about a week, when solar wind streams are especially fast. Whereas satellite encounters with the SAA are as predictable as the tides, usually solar energetic particles, geomagnetic storms, and high-speed solar wind streams are not reliably predictable, nor is the intensity of the associated radiation event. The high-inclination orbit of ISS therefore introduces a new radiation risk factor.

ISS construction plans call for approximately 33 U.S. shuttle flights and 10 Russian flights. The construction phase will extend from 1998 to 2004, which spans the maximum of solar cycle 23, when SPEs are expected to be most frequent (see Figure 1.4). NASA estimates that during those years astronaut and cosmonaut construction crews may have to perform more than 160 EVAs totaling more than 1,100 hours. During those same years, there will be more than 400 additional hours of EVAs by astronauts and cosmonauts to service and maintain the station. The total exceeds 1,500 hours, or 1,000 ISS orbits, of EVA time.

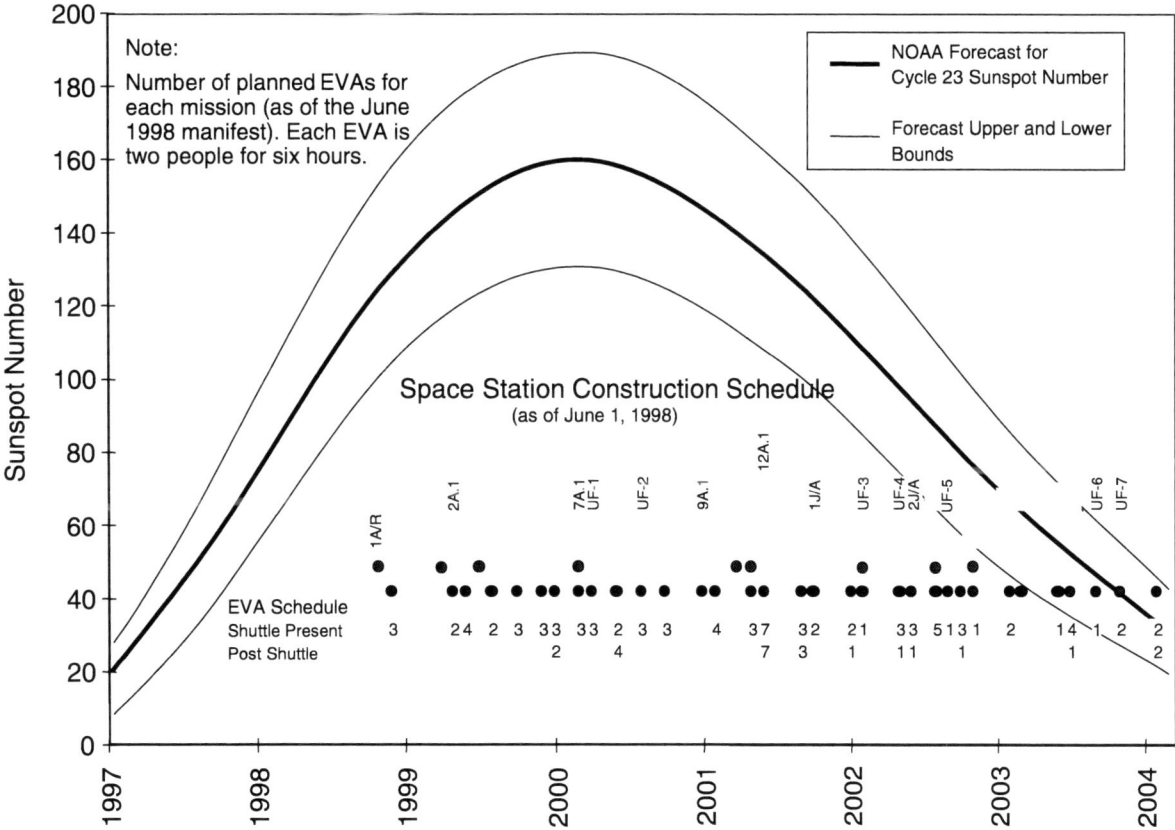

Figure 1.4 Solar cycle 23 and EVA schedule.[19]

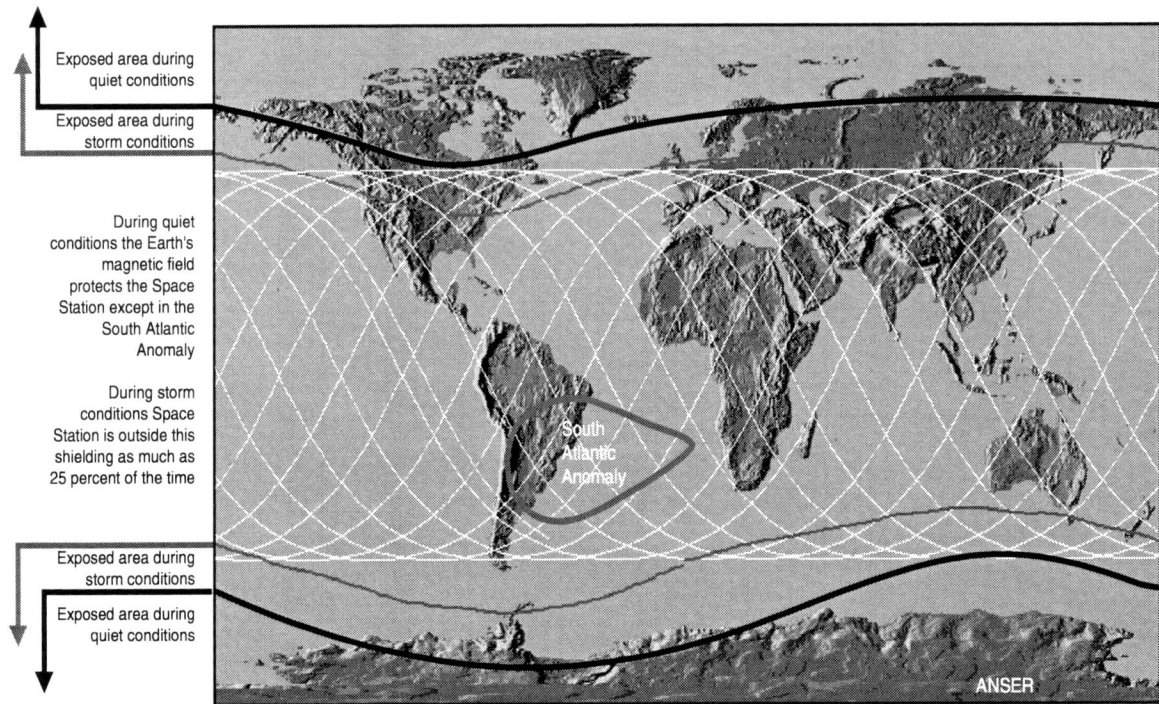

Figure 1.5 Twenty-four hours of ISS ground track overlaid with the magnetic shielding boundaries for quiet and active storm conditions and the SAA. The quasi-latitudinal pair of high-latitude lines in each hemisphere indicates the low-latitude borders of areas accessible to radiation from SPE zones. The higher-latitude line in each hemisphere represents quiet conditions and the lower-latitude line, disturbed conditions.[20]

Figure 1.5 illustrates the geometry relevant to discussing the risk factors faced by astronauts and cosmonauts during EVAs. It traces the sixteen 90-minute orbits ISS will make each day and shows the approximate border of the SAA. The borders of the polar caps to which solar energetic particles have access are the high-latitude curves, whose troughs and crests reflect the 11.4-degree tilt of the magnetic dipole axis relative to the geographic axis. This high-latitude pair of curves delineates the approximate areas to which solar energetic particles have access during magnetically quiet times. The lower-latitude pair of curves delineates the approximate areas to which solar energetic particles have access during severe geomagnetic storms. Severe SPEs and severe geomagnetic storms have a common cause (CMEs) and often overlap in time (see Section 2.3). The lower latitude pair of borders defines the areas for assessing the SPE risk to astronauts during EVAs in ISS orbit—the areas to which SPE particles have direct access are called the SPE zones. Cosmic ray physicists refer to the equatorial border of the SPE zones as the cosmic ray cutoff.

Figure 1.5 reveals three groups of orbits that intersect radiation zones. One group crosses the SAA on the descending leg of the orbits, a second group crosses it on the ascending leg of the orbits, and a third group crosses the SPE zones. The figure also shows that as the ISS moves along its 16 orbits per day, its per-orbit risk of exposure to solar energetic particles varies from zero to a maximum that depends on the size of the SPE zones.

Figure 1.6 quantifies the per-orbit SPE exposure risk over 24 hours as Earth rotates the "off-center" SPE zones toward and away from the ISS orbit. The rotation causes the SPE zones to overlap ISS orbit for a portion of each day by an amount that depends on the size of the zones at the time. The orbit's position is specified by the

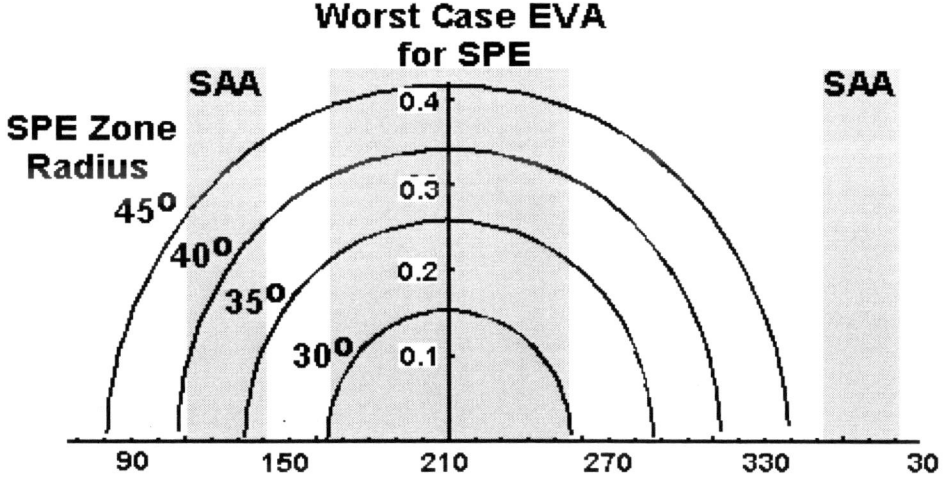

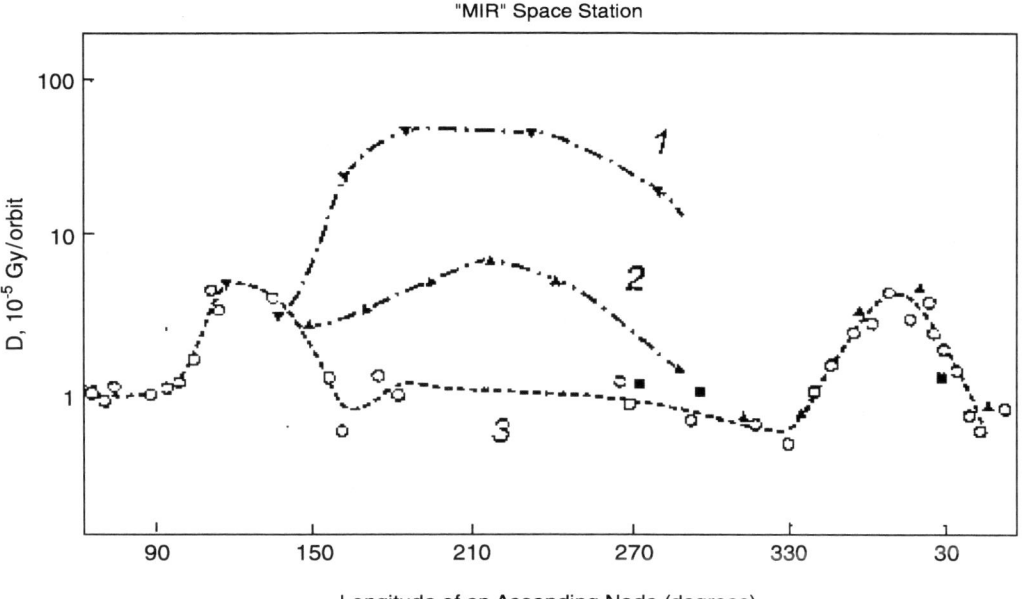

Figure 1.6 SPE and SAA zones as a function of the longitude of the ascending node of the ISS and Mir orbits. Shading in the top panel shows the locations of the SPE and SAA zones. Circles show the fraction of the orbit inside SPE zones for four assumed radii of the zone (30 to 45 degrees in 5-degree steps) based on simplified geometry. The bottom panel shows radiation data from the Mir space station during a non-SPE orbit (line 3) and two orbits during the SPE of September 29, 1989.[21]

longitude of its ascending node, that is, the point at which it crosses the equatorial plane from south to north. The SPE exposure risk is given as the fraction of the orbit that overlaps the SPE zones. The fraction of overlap is calculated using a simple model of SPE-zone geometry in which a circle centered on the axis of Earth's magnetic dipole axis represents the zones. The axis is taken to pass through Earth's center and to penetrate Earth's surface in the northern hemisphere at 78.5 degrees latitude and 69 degrees west longitude. The model ignores the offset of the dipole axis from Earth's center (which gives rise to the SAA but does not much affect the position of the area accessible to SPE particles), and it ignores the antisolar shift of the zones, which varies with geomagnetic activity and can be several degrees.[22] The antisolar shift of the zones causes the fraction of overlap to exceed the calculated value for a given orbit in one hemisphere and to lag it in the other, so that the effect tends to cancel out on a per-orbit basis. Figure 1.6 presents results for 30-, 35-, 40-, and 45-degree SPE zones, which cover conditions from typical to extreme. As an example, consider the worst case. Of the 16 ISS orbits per day, the orbit whose longitude of ascending node is nearest 210 degrees overlaps the SPE zones the most (refer to Figure 1.5). If while ISS is traversing one of these worst-case orbits the SPE zones happen to be 30 degrees in radius, corresponding to quiet conditions, ISS will be in the zones about 15 percent of the time. If, on the other hand, conditions happen to be highly disturbed, so that the zones are 45 degrees in radius, the percentage increases to 42. To estimate the worst-case fraction for an entire EVA, however, one must average the fraction per orbit over four orbits—corresponding to the wide shaded area in Figure 1.6—since EVAs nominally last 6 hours but usually extend to 7 hours (plan 6, do 7). Then, in the worst case, the average fraction per orbit spent in the SPE zones varies from about 8 percent for the 30-degree zone to about 40 percent for the 45-degree zone. (The estimate ignores the quantization introduced by discrete orbits of 90 minutes duration.) The two narrow shaded regions in Figure 1.6 show where the ISS orbit intersects the SAA twice per day on its ascending and descending swings.

The bottom panel in Figure 1.6 gives an example from the Mir space station of per-orbit dose data that illustrate the information given in the top panel. The recording instrument, which was inside the station behind an average of 10 to 15 g/cm^2 of shielding (compared with about 0.5 g/cm^2 for a space suit), was sensitive to protons with energies above 100 MeV. Line 3 shows a non-SPE situation in which the two SAA peaks stand out as order-of-magnitude increases in the dose rate. Lines 1 and 2 show per-orbit doses during, respectively, the peak of an SPE on September 29, 1989, and three days later. The shape of line 1 suggests that the effective radius of the SPE zones at this time was between 30 and 35 degrees, since the dose per orbit drops substantially around 150 and 270 degrees longitude of ascending node.

To predict the severity of the SPE radiation risk, one also needs to know the dose rates that would represent a worst-case scenario. The great solar particle event of August 4, 1972, is a commonly used benchmark for worst-case estimates. During that event, the solar proton detectors on the spacecraft monitoring SPE fluxes became saturated, so the value of the peak flux is uncertain. This uncertainty has resulted in estimates of peak dose rate that vary by a factor of two. Two studies in this area are those of Letaw et al.[23] and Wilson et al.[24] Letaw et al. put the peak dose rate to the blood-forming organs (BFO) of an astronaut in a space suit at about 0.3 Sv/h, whereas Wilson et al. put it at approximately 0.15 Sv/h. Letaw et al. also give a worst-case composite dose rate of about 0.5 Sv/h. These rates are assumed to apply to the SPE zones. The peak dose rate lasted about 8 hours, which gives a total dose of between 1.2 Sv and 2.4 Sv, depending on which dose-rate estimate is used. For comparison, a major SPE on October 19, 1989, is estimated to have produced a total BFO dose behind EVA-level shielding of 1.29 Sv.[25] This is similar to the Wilson et al. estimate for the 1972 event, although the time histories of the two events are not similar.

The difference between the two dose-rate estimates can be traced to different dose-rate protocols and to the sensitivity that these protocols have to aspects of the particle spectrum and composition that were not adequately measured during the August 1972 event. The Wilson et al. number,[26] being the latest and so built on prior estimates, gives the current best estimate. But it should be noted that the procedure for calculating the dose rate is still somewhat fluid. That is, there is more than one protocol for calculating dose rates, and the results using different protocols do not always agree. A separate point of disagreement concerns the value of constructing a composite worst-case estimate. The Letaw et al.[27] composite worst case is similar to an envelope over the spectra from all worst cases. It combines the worst case for low-energy fluxes (the 1972 event) with the worst case for high-energy fluxes (a 1956 event). Some radiation scientists consider this artificial "SPE from Hell" unduly extreme, since no single event is likely to exhibit the worst features of all prior events. On the other hand, it is also unlikely that the maximum dose rate of all prior events will never be exceeded. CSSP/CSTR therefore used all three estimates in constructing Table 1.3. The lowest values came from Wilson et al. and should be regarded as the current best worst-case estimates. The intermediate values came from Letaw et al.[28] and should be taken as representing the uncertainty in constructing a worst case from the August 1972 event. The highest values came from the Letaw et al.[29] composite and should be considered not a worst case but a conservative upper limit. For the sake of making a worst-case estimate for each dose rate, CSSP/CSTR assumed that the EVA occurred during the peak of the 1972 event and that it was centered on the maximum excursion into the SPE zones (i.e., the wide gray area in Figure 1.6).

Table 1.3 gives the total dose based on different SPE zone radii and different dose rate assumptions. These numbers can be compared with an upper limit of 0.15 Sv estimated for the dose to the Mir cosmonauts inside the station during the October 1989 SPE.[30] This upper limit estimate should be multiplied by 3.5 to be applicable to an EVA situation, as determined by direct comparison of the dose rates inside and outside the station (Gautam Badhwar, SRAG, JSC, private communication, 1999).

The lower range of worst-case estimates (first row of Table 1.3) encroaches on an astronaut's short-term radiation limits (Table 1.1). The upper range (second row) encroaches on an astronaut's career radiation limit (Table 1.2). The high end of conservative upper-limit values (third row) approaches and somewhat overlaps the range in which symptoms of acute radiation sickness begin to occur. A similar table for estimates of worst-case doses to the skin and ocular lens would show that the short-term limits are reached at doses lower by a factor of about two.[31]

Of course, by definition, worst-case scenarios are unlikely to happen. It is unlikely, for example, that the peak of the SPE dose rate would coincide with a 6-hour EVA. But if the construction schedule calls for consecutive EVAs, one EVA shift might receive a dose of about 50 percent or more of the total possible. As another example, it is unlikely that an SPE will occur during any given construction flight. But because ISS requires a large number of construction flights and because the construction will occur during the peak of solar cycle 23, the probability is high that some construction flights will experience SPE radiation. Section 2.2 of this report presents a calculation showing that for specified assumptions about the severity of solar cycle 23, the probability that at least two flights will experience SPE conditions is close to 100 percent. The probability drops to about 50 percent for five to seven flights experiencing SPE conditions. Section 2.3 shows that SPEs with high dose rates tend to occur when the SPE zones are big, 35 degrees or more. Chapter 2 therefore reaches an important conclusion: there is a nonnegligible likelihood that while performing an EVA during the construction phase of ISS, astronauts could receive a radiation dose that is significant in terms of increased cancer risk and of reaching allowable dose limits. The likelihood of this happening could diminish, however, if the ability of NASA to specify and forecast SPE severity at the ISS were to improve.

The analysis presented here deals with the radiation dose received during a single SPE. One must add this dose to the accumulated dose that an astronaut receives from all sources during a mission and from prior missions.

Table 1.3 Worst-Case Estimates of BFO Dose for a 6-Hour EVA with Maximum Exposure Within SPE Zones Having Radii and Dose Rates As Shown

Dose Rate (Sv/hr)	Dose (Sv)			
	30°	35°	40°	45°
0.15	0.06	0.20	0.29	0.36
0.30	0.13	0.40	0.58	0.72
0.50	0.21	0.66	0.96	1.20

NOTE: Space suit shielding is assumed to be equivalent to about 0.5 g/cm^2 Al.

Insofar as it would be more likely to cross a radiation limit threshold, the incremental dose that an SPE delivers toward the end of a mission could be more serious than a dose delivered in the early part of a mission.

Figure 1.6 illustrates a fairly obvious but operationally very important point: if a flight director knows that a serious SPE is in progress or will be in progress and has the flexibility to change preset EVA schedules, then scheduling EVAs that avoid deep penetration of the SPE zones, even if they straddle the SAA, would greatly reduce the particle radiation dose to astronauts. This point demonstrates the value of employing data and implementing models that reliably specify and predict SPE occurrence, intensity, and duration. Such data are discussed in Chapters 3 and 4, and such models are discussed in Appendix A.

1.5 ISSUES IN MANAGING RADIATION RISK DURING ISS CONSTRUCTION

The flight director at JSC has overall responsibility for the safe execution of a mission. Radiation risk is only one issue the flight director must consider in deciding whether to delay a launch or an EVA or whether to end a flight or an EVA early. In some construction activities, the negative impact on construction logistics of delaying an EVA or ending it early because of an SPE could exceed the negative impact of the additional dose the astronauts would receive by keeping to the original schedule. NASA has accepted the radiation dose limits recommended in the 1989 NCRP report.[32] It is likely to accept the new NCRP recommendation that calls for reducing the career limit by one half (Table 1.2). In general, these limits are applied retrospectively, that is, after a flight. During the flight, the primary operative flight rule is to keep the dose as low as reasonably achievable (ALARA). This rule gives the flight director much discretion. For example, it does not preclude an astronaut's reaching his or her 30-day or annual limit during one flight, although this has never happened. If the SPE dose rate should become high enough and the SPE zones become wide enough to threaten doses in the career-limit range or in the acute radiation sickness range during an EVA, then obviously radiation would become a bigger component in the risk equation that the flight director must consider. To factor this component into the risk equation, however, the flight director must have sufficiently reliable information about the radiation environment. Part of this report has to do with assessing which information is now or can become sufficiently reliable.

An unofficial flight rule, but one flight directors nonetheless observe, is the real-time, on-site flight rule, which says that any decision made in response to a radiation situation must be based on radiation measurements made in real time and on site, that is, in the external ISS environment. This requirement cannot be fulfilled if current plans are followed, because from now until June of 2000 there will be no capability for obtaining such measurements, and only retrospective dose data will be available. As already stated, this report discloses a high probability that a flight director will face an SPE-radiation situation during ISS construction and that the dose increments during an EVA while an SPE is in progress can account for a significant fraction of an astronaut's short-term limits or even exceed those limits.

Whatever a flight director decides in response to a radiation situation, the primary input to that decision pertaining to astronaut health and safety comes in the form of a recommendation from the flight surgeon.[33] The flight surgeon, who knows the astronauts' radiation histories, can tell how much an incremental dose of radiation advances each astronaut toward his or her radiation limits. To make a real-time recommendation to the flight director to alter an EVA schedule or assignment based on a radiation situation, the flight surgeon needs a complete and accurate description of the real-time, on-site radiation situation.

The unit at JSC responsible for informing the flight surgeon about radiation situations is SRAG. This unit has multiple responsibilities: provide preflight crew exposure projections; provide real-time astronaut radiation protection support; provide radiation monitoring to meet medical and legal requirements; maintain comprehensive crew exposure modeling capability; provide preflight planning and analysis support; and provide in-flight support. In-flight support entails specific responsibilities, among them the following: provide updated EVA exposure analysis to the flight surgeon; provide EVA start and stop times to the flight surgeon; provide an EVA go/no go recommendation prior to egress; monitor real-time space weather; recommend whether to continue or terminate an EVA during a radiation event; track exposure from the nominal radiation environment; monitor extravehicular charged-particle directional spectrometer data when they become available starting in June of 2000; and provide a post-EVA final estimate of additional crew exposure to the flight surgeon and the flight director. SRAG consists of a small number of health physicists, physicists, and programmers. It has one civil servant (or possibly none) and four or five contractors. Considering the number of scheduled EVAs during ISS construction, it may be overtasked.

SRAG has available a number of resources with which to execute these functions: statistical radiation-belt models, models for specifying and predicting SPE conditions, real-time data from SEC on the space weather and the radiation environment, and codes to evaluate the radiation situation at ISS from the statistical models and the real-time data. Equipment for monitoring the ISS radiation environment comprises crew passive dosimeters (CPDs) and a tissue-equivalent proportional counter (TEPC). CPDs provide postflight data. The TEPC gives dose-rate and cumulative-dose data every several seconds to the crew, but the data are not telemetered to SRAG. TEPC data refer to the radiation environment inside the shuttle or station, not to the EVA environment. An external radiation monitor is scheduled to be mounted on the station during the eighth shuttle flight, in June of 2000, after 46 ISS EVAs will already have been performed.

Although SRAG uses state-of-the-art models, the (unofficial) flight rule excluding nonlocal, non-real-time data suggests that such data specify the radiation environment at a confidence level too low to allow convincing a flight director to change a flight schedule or an EVA schedule in the absence of on-site data. This implies that SRAG would be more effective if its models could achieve a confidence level high enough that a flight director would consider the model's specifications as adequate substitutes for on-site data.

Astronauts are the stakeholders most directly affected by how radiation risk is managed. A heavy radiation dose while performing an EVA during an SPE puts an astronaut in double jeopardy: no astronaut wants to reach the short-term radiation limits, much less the career limit. Reaching a radiation limit is the first jeopardy. The second is the increased risk of stochastic radiation effects, that is, induced cancer. This jeopardy does not go away after the flight or even after the career.

There is also another party involved in radiation risk management that is not directly involved in ISS construction. This is the community that studies space weather and provides space weather services. The space weather community is made up of physicists at universities, government laboratories, and industry who study the environment of space and researchers and forecasters at space weather operation centers. This community currently possesses resources and capabilities that could be marshaled to develop tools to specify and forecast the radiation environment at ISS orbit more completely and accurately than is now done.

1.6 THE APOLLO EXPERIENCE

During the Apollo program a network of solar observatories was used to give a real-time warning of SPEs and to estimate their possible impact on the lunar missions. The Solar Particle Alert Network (SPAN) was implemented by the NASA Manned Spacecraft Center (MSC)—now the Johnson Space Center—in Houston.[34] It consisted of seven observatories located around the world to ensure 24-hour observations of solar activity. Hydrogen-alpha (0.5 Å bandwidth) telescopes were installed at all seven observatories to observe optical solar flares that produce solar particle events. The observatories were located at Houston; Boulder, Colorado; Honolulu, Hawaii; Carnarvon, Australia; Culgoora, Australia; Teheran, Iran; and the Grand Canary Islands, Spain. Each location had real-time communications with Mission Control Center (MCC) in Houston. In addition, there were radio-frequency telescopes operating at 2695 MHz at three locations: Houston, Carnarvon, and the Canary Islands. A blue-ribbon committee of space scientists, chaired by Wilmot Hess, oversaw the implementation of the network.

It was recognized early in the Apollo program that high-energy particles from solar flares could pose a radiation hazard to the astronauts. They were especially vulnerable while they were in the thinly shielded lunar excursion module (LEM) or on the lunar surface. However, the command and service module provided enough protection to reduce exposures from solar particle events to acceptable levels. The Apollo missions were scheduled to take place during solar maximum years, when large solar particle events are more apt to occur. Research had established that virtually all particle events during solar cycle 19 were preceded by type IV solar radio bursts. However, not all type IV bursts were followed by particle events. (The same is true for solar flares observed in the hydrogen-alpha line, but there are many more flares than type IV radio bursts.) Studies carried out at MSC established a correlation between large type IV solar radio bursts and SPE size (time-integrated proton flux

> 30 MeV). The radio flux was integrated over time to obtain a measure of the energy of the burst. The hypothesis was that the radio burst was produced by synchrotron radiation from electrons that are accelerated at the same time as the protons. Data from radio observatories at Ottawa, Canada (which operated at 2800 MHz) and Nagoya University, in Japan (which operated at 3000 MHz) were used for the study. Particle event data were taken from the Solar Proton Manual and a Boeing Company report.[35]

While some solar flares produce relativistic-energy protons that can arrive in the Earth-Moon region within 30 minutes, the arrival times for most events are 4 to 6 hours after the flare and radio burst. Peak particle intensities do not occur until another 4 to 6 hours after the arrival of particles. The strategy was to use this time to move the Apollo astronauts off the lunar surface and have them return to the more heavily shielded command and service module. Information on the occurrence of a solar flare (observed by the hydrogen-alpha telescopes) and data from a large radio-frequency (10 cm wavelength) burst were transmitted back to MCC in Houston. Radiation specialists working on the radiation console, located in one of the MCC "back rooms," analyzed these data. If an event of a certain (estimated) size was believed to produce a substantial radiation dose to the astronauts, the flight director would be advised so that action could be taken to minimize their exposure. Particle spectrometers and dosimeters onboard the Apollo spacecraft then detected the increase in the radiation environment, to verify that the particle event had arrived in cislunar space. This reduced the impact of a false alarm when a flare and type IV burst did not produce particles that propagated to cislunar space. Flight rules precluded launching into an SPE or landing on the Moon during such an event and terminated a lunar excursion if exposures were estimated to be above an acceptable level.

Fortunately, no large SPEs occurred during any Apollo mission. The large solar particle event of August 1972 occurred between the Apollo 16 and Apollo 17 missions and did not, therefore, affect them. A solar flare and radio burst occurred during the Apollo 12 mission, which had exercised the operational procedures.

NASA operated SPAN until late 1970, after Congress gave NOAA responsibility for space weather. NOAA tasked the Space Environment Forecast Center in Boulder with carrying out this responsibility, and NASA transferred SPAN to NOAA, which operated it throughout the remaining Apollo flights and during the Skylab missions. Although NOAA continues to participate in the space radiation protection program for shuttle missions, it has replaced SPAN with other observational methods, including spaceborne X-ray and particle instruments.

It is of interest that the Soviet space program used similar SPE warning criteria throughout the 1970s and 1980s (Vladislav M. Petrov, Institute for Biomedical Problems, Moscow, personal communication). Russians performing EVAs at ISS will be directed out of the mission control center in Moscow. Further, it is likely that U.S. and international crew members on ISS will also participate in EVAs directed out of MCC-Moscow, whose flight rules pertaining to radiation may differ from those of MCC-Houston. While this report is focused on U.S. policy regarding radiation risk and ISS, CSSP/CSTR believes some of the recommendations in this report might also be implementable by MCC-Moscow.

1.7 SUMMARY AND RECOMMENDATION

ISS construction and concurrent station maintenance will entail more than 1,500 hours of EVAs over a 4-year period, 1998 to 2002, that straddles the peak in the current solar cycle. The station's high-inclination orbit (51.6 degrees) cuts through radiation environments more severe than those of the originally planned low-inclination orbit (28 degrees). The high-latitude radiation environments (energetic particles from solar storms and relativistic electrons in Earth's outer radiation belt) are highly variable. At the height of their variability, they are intense enough to pose a hazard to astronauts engaged in EVAs, although even doses estimated for worst-case scenarios fall short of life-threatening. (This is in contrast to the situation astronauts could face in flights beyond the protective shield of Earth's magnetic field—for example, on a flight to Mars.[36])

Although the fraction of time during which the high-latitude radiation environments reach threatening levels is small, the amount of time committed to EVAs is large, so the probability that an SPE will occur when an EVA is scheduled is not small. Estimates based on calculations referred to in this report put at near certainty the likelihood that at least two ISS construction flights will be in progress when energetic particles from a solar storm invade a volume of space above the polar atmosphere that at times of solar storms often significantly overlaps the ISS orbit.

Flight directors will almost certainly be faced with flights impacted by high-latitude radiation. This finding directs attention to the issue of radiation risk management. The responsibilities at JSC in this area are well defined. The CSSP/CSTR review of current JSC flight rules shows, however, that radiation risk management would benefit by exploiting resources and capabilities that currently exist elsewhere. This report discusses the relevant resources and capabilities, addresses issues related to their application to radiation risk management, and makes recommendations aimed at reducing radiation risk. One main recommendation emerges from Chapter 1.

Recommendation 1: Because it denies access to valid information and thus unnecessarily restrains flight-director options, flight directors should not adhere rigidly to the (unofficial) real-time, on-site data rule.

Simply put, reducing the dose to astronauts from solar energetic particles during EVAs entails avoiding EVAs during orbits that penetrate the SPE zones when SPE particles are present. To implement an operational procedure for SPE-zone avoidance, the flight director must act on information on the size and shape of the zones and on the occurrence, intensity, and duration of the SPE. Information on SPE zone geometry and SPE start, strength, and length must be reliable enough to gain the flight director's confidence and timely enough to allow the flight director to act on it. The status of the resources (existing or in development) that are needed to acquire such information is reviewed in Chapters 3 and 4 and Appendix A.

One might draw an analogy to the influence of terrestrial weather in the execution of space missions. A flight director will routinely delay a launch or a landing because of a thunderstorm forecast based on nonlocal data. Forecasts of space weather based on nonlocal data could, likewise, help the flight director reduce the probability of radiation exposure to astronauts during EVAs. The principal difference in these two cases is that a terrestrial weather forecast has consequences for hardware, whereas a space-weather (radiation) forecast affects the rotation of astronauts and their health and future flight opportunities.

For Recommendation 1 to be successful, flight directors should acquire ownership of any flight rule that replaces the one whose removal is recommended. They should therefore work with SRAG to approve sources of nonlocal data and models that use these data to specify and forecast radiation levels at ISS. Radiation conditions at ISS should be inferred from nonlocal data until resources become available to augment the nonlocal data and models with real-time, on-site data and modeling.

One instance indicates that flight directors might be able to participate more in reducing radiation risk. SRAG proposed putting a radiation monitor in the shuttle bay to provide real-time, on-site radiation data on flights before Flight 8A, which will install a radiation monitor on the station. However, the proposal failed to receive flight director approval in time to implement it before Flight 8A. This delay guaranteed that, under current operating procedures, no change to flight or EVA schedules in response to a radiation situation is possible prior to Flight 8A. Timely action by the flight directors could have made it possible to respond to a radiation situation under current flight rules.

1.8 NOTES AND REFERENCES

1. The 1999 report on space weather of the National Security Space Architect finds that during the preceding 16 years at least 13 satellites suffered total mission failure attributable to space weather. There were more failures in which space weather was implicated, but the evidence was not definitive.

2. J.R. Letaw, R. Silberberg, and C.H. Tsao, "Galactic cosmic radiation doses to astronauts outside the magnetosphere," in *Terrestrial Space Radiation and its Biological Effects*, P.D. McCormack, C.E. Swenberg, and H. Bucker, eds., Plenum Press, New York, 1988; J.W. Wilson, F.A. Cucinotta, J.L. Shinn, L.C. Simonsen, R.R. Dubey, W.R. Jordan, T.D. Jones, C.K. Chang, and M.Y. Kim, "Shielding from solar particle event exposures in deep space," in *Proceedings of Workshop on Impact of Solar Energetic Particle Events for Design of Human Missions*, September 9-11, 1997, Center for Advanced Space Studies, Houston.

3. Construction of the International Space Station, a project of the United States (lead), Canada, Japan, the European Space Agency, and the Russian Federation, began in late 1998. The ISS is in orbit at an altitude of 250 statute miles with an inclination of 51.6 degrees. The first crew to live aboard the (partially assembled) space station is scheduled to arrive in March 2000. Assembly of the ISS is scheduled to continue until late 2004.

4. D.N. Baker, "Solar wind-magnetosphere drivers of space weather," *J. Atmos. Terr. Phys.*, 58, pp. 1509-1526.

5. National Research Council, Committee on Solar and Space Physics and Committee on Solar-Terrestrial Research, *Space Weather: A Research Perspective*, available on the Internet at <www.nas.edu/ssb/cover.html>.

6. See, for example, J.F. Lemaire, D. Heynderickx, and D.N. Baker, eds., *Radiation Belt Models: Models and Standards, Geophysical Monograph 97*, Washington, D.C.: American Geophysical Union, 1996. The SAA has been monitored in detail sufficient to follow its westward drift of roughly 1 degree every 3 years owing to the secular variation of the geomagnetic field. G.D. Badhwar, "Drift rate of the South Atlantic Anomaly," *J. Geophys. Res.*, 102, 1997, pp. 2343-2349.

7. The National Research Council has produced a document on this topic (see footnote 5), which may be consulted for a fuller treatment and more detail.

8. Definition taken from the Strategic Plan of the National Space Weather Program, 1995, obtainable from the Upper Atmospheric Section of the Division of Atmospheric Sciences, National Science Foundation.

9. R.A. Howard, M.J. Koomen, D.J. Michels, R. Tousey, C.R. Detwiler, D.E. Roberts, R.T. Seal, and J.D. Whitney (U.S. Naval Research Laboratory, Washington, D.C.) and R.T. Hansen, S.F. Hansen, C.J. Garcia, and E. Yasukawa (High Altitude Observatory, NCAR, Boulder, Colo.), "Synoptic observations of the solar corona during Carrington rotations," 11 October 1971-15 January 1973, pp. 1580-1596 (Reissue of UAG-48 with quality images, February 1976, 200 pp. Supersedes UAG-48).

10. R.M. MacQueen, J.A. Eddy, J.T. Gosling, E. Hildner, R.H. Munro, G.A. Newkirk, A.I. Poland, and C.L. Ross, "The outer corona as observed from Skylab," *Astrophys. J.*, 187, 1974, p. L85.

11. For more information about our current understanding of CMEs and their relation to other manifestations of solar and geomagnetic activity, see N.U. Crooker, J.A. Joselyn, and J. Feynman, eds., *Coronal Mass Ejections, Geophys. Monogr. Ser.*, 99, American Geophysical Union, Washington, D.C., 1997.

12. S. Kahler, "Solar flares and coronal mass ejections," *Ann. Rev. Astron. Astrophys.*, 30, 1992, pp. 113-141.

13. J.T. Gosling, "The solar flare myth," *J. Geophys. Res.*, 98, 1993, p. 18937.

14. Gautam Badhwar, Presentation to CSSP/CSTR. This material, which was presented to CSSP at its meeting on January 26, 1998, is available for viewing in the National Research Council's Public Access Records Office.

15. National Council on Radiation Protection and Measurement, *Guidance on Radiation Received in Space Activities*, Report No. 98, 1989, p. 70; R.W. Young, "Acute radiation syndrome," in *Military Radiobiology*, J.J. Conklin and R.I. Walker, eds., Academic Press, New York, 1987.

16. National Council on Radiation Protection and Measurement, *Guidance on Radiation Received in Space Activities*, Report No. 98, 1989, p. 71.

17. National Council on Radiation Protection and Measurement, *Guidance on Radiation Received in Space Activities*, Report No. 98, 1989, p. 73.

18. National Council on Radiation Protection and Measurement, *Guidance on Radiation Received in Space Activities*, Report No. 98, 1989.

19. Figure from R. Turner and C. Kemere, "Solar particle events and International Space Station," Report submitted to the Committee on Solar and Space Physics, August 12, 1998.

20. The ground track was computed by the Analytical Graphics Incorporated (AGI) Satellite Tool Kit from the recent ISS ephemeris. The magnetic shielding boundaries are 30 MeV geomagnetic vertical cutoff calculations by Don Smart for Kp = 0 (quiet) and Kp = 9+ (active), from D.F. Smart, M.A. Shea, E.O. Flueckiger, A.J. Tylka, and P.R. Boberg, *Changes in Calculated Vertical Cutoff Rigidities at the Altitude of the International Space Station As a Function of Geomagnetic Activity*, 26th International Cosmic Ray Conference, Contributed Papers, Vol. 7, 1999, pp. 337-340. The South Atlantic Anomaly boundary is for 100 MeV protons with flux greater than 100 particles/cm^2-sec, as calculated by C. Dyer, A. Sims, and C. Underwood using the NASA AP-8 model with a 1991 magnetic field model, in "Radiation belt observations from CREAM and CREDO," Geophysical Monograph 97, *Radiation Belts: Models and Standards*, J.F. Lemaire, D. Heynderickx, and D.N. Baker, eds., American Geophysical Union, Washington, D.C., 1996. The compilation was produced by Ron Turner and Stephen Thomas, ANSER, Arlington, Va.

21. V.A. Shurshakov et al., "Solar particle events observed on Mir station," in *Proceedings of Workshop on Impact of Solar Energetic Particle Events for Design of Human Missions, September 9-11, 1997*, Center for Advanced Space Studies, Houston, 1998, pp. 1-18.

22. G.L. Siscoe, "What determines the size of the auroral oval?" in *Auroral Physics*, C.-I. Meng, M. Rycroft, and L.A. Frank, eds., Cambridge University Press, 1991, pp. 159-175.

23. See Note 2.

24. See Note 2.

25. J.T. Lett, W. Atwell, and M.J. Golightly, "Radiation hazards to humans in deep space: A summary with special reference to large solar particle events," in *Solar-Terrestrial Predictions, Proceedings of a Workshop at Leura, Australia, October 16-20, 1989, Vol. 1*, NOAA/ERL, 1990, pp. 140-153.

26. See Note 2.

27. See Note 2.

28. See Note 2.

29. See Note 2.

30. G. Badhwar, Presentation to CSSP. This material, which was presented to CSSP at its meeting on January 26, 1998, is available for viewing in the National Research Council's Public Access Records Office.

31. J. Wilson, L. Townsend, W. Schimmerling, G. Khandelwal, F. Khan, J. Nealy, F. Cucinota, L. Simonsen, J. Shinn, and J. Norbury, *Transport Methods and Interactions for Space Radiation*, NASA Reference Publication 1257, 1991.

32. See pp. 7 and 8 in the reference at note 16.

33. A proposed flight rule, submitted to the ISS program office for approval, also gives a prominent role in real-time decision making during ISS missions to the radiation health officer (RHO). Under this rule, the flight surgeon would consult with the RHO regarding potential radiation effects (based on radiation dose projections from SRAG) before making recommendations to the flight director regarding actions to reduce crew radiation exposure. Such a consultation might occur, for example, as astronauts prepare to leave the relative safety of the ISS for an EVA, or while an EVA is under way.

34. J.L. Modisette, M.D. Lopez, and J.W. Snyder, "Radiation plan for the Apollo lunar mission," AIAA paper 69-19; J.L. Modisette, T.M. Vinson, and A.C. Hardy, "Model solar proton environments for manned spacecraft design," Manned Spacecraft Center, NASA TN D-2746, April 1965; M.D. Lopez, A.L. Bragg, and J.L. Modisette, "Preliminary warning criteria for Solar Particle Alert Network," NASA Program Apollo Working Paper No. 1193, NASA Manned Spacecraft Center, 1966; D.E. Robbins and J.H. Reid, "Solar physics at the NASA Manned Spacecraft Center," *Solar Physics*, 10, 1969, pp. 502-510.

35. H.H. Malitson and W.R. Weber, *Solar Proton Manual*, F.G. McDonald, ed., NASA Goddard Space Flight Center X-611-62-122, 1963; W.R. Weber, "An evaluation of the radiation hazard due to solar-particle events," Boeing Co. Report D2-90469, December 1963.

36. Space Studies Board, National Research Council, *Radiation Hazards to Crews of Interplanetary Missions*, National Academy Press, Washington, D.C., 1996, p. 15.

2

Solar Particle Events and the International Space Station

2.1 BACKGROUND TO AN ASSESSMENT OF SPE IMPACTS ON ISS CONSTRUCTION

Solar storms often accelerate ions to energies that can penetrate space suits and even spacecraft. Occasions when this occurs are often called solar particle events (SPEs) but may also be called solar cosmic ray events, solar proton events, solar energetic particle events, energetic storm particle events, ground-level events, proton showers, or polar cap absorption events. Many of these terms, which are for most purposes synonymous, are still in use in the scientific literature. In this report, CSSP/CSTR consistently used "solar particle event" and its abbreviation, SPE.

NOAA's Space Environment Center (SEC) declares an SPE to be in progress when the dose rate of particles with energies above 10 MeV (i.e., space-suit-penetrating) exceeds 10 particles $cm^{-2}s^{-1}sr^{-1}$ (directional flux) for more than 15 minutes. When this happens, SEC alerts the Space Radiation Analysis Group (SRAG) at the Johnson Space Center (JSC), recalling SRAG to Mission Control Center (MCC). Should the dose rate of particles with energies above 100 MeV exceed 1 particle $cm^{-2}s^{-1}sr^{-1}$, an "energetic SPE" is declared, which mandates that SRAG remain on console. Such events can last several days.

Two criteria can be used to define when a flight is significantly impacted by an SPE. Criterion 1 says that a flight is impacted if an SPE occurs that reaches the "alert" (10 particles $cm^{-2}s^{-1}sr^{-1}$) stage. Criterion 2—the "significant dose" criterion—says that it is impacted if an SPE occurs with an accumulated free-space dose of 10^8 particles cm^{-2} (omnidirectional fluence) and energies above 10 MeV. Converted into a dose to tissue, this corresponds to about 0.6 Gy (see Section 1.3 for definitions of radiation units),[1] the actual number depending on the energy spectrum of the SPE particles. The second, more stringent criterion marks a condition that in the worst-case orbit geometry relative to the SPE zones and worst-case EVA timing, would noticeably increase an astronaut's radiation dose. Criterion 1 can be met without criterion 2 being met, but rarely if ever can the reverse happen. When criterion 1 is met, the flight's ground support personnel must be placed in a state of radiation alert, affecting decisions on when to launch, which astronauts to assign to which tasks, EVA schedules, and when to return. When criterion 2 is met, the repercussions for a flight and its crew are more serious, because a radiation condition has occurred that could bring an astronaut significantly closer to the specified limits for personal radiation exposure.

The science of SPE phenomenology, upon which radiation risk assessments like those described below are based, started over 50 years ago. It was an offshoot of cosmic ray research, which began even earlier, after Victor Hess demonstrated that there was a constant flux of ionizing radiation entering the atmosphere from space. In 1925, Robert A. Millikan dubbed this radiation "cosmic rays." The study of cosmic rays blossomed in the early

1930s, when the invention of the ionization chamber allowed more sensitive experiments. Now we know that there are two kinds of cosmic rays, galactic and solar. Galactic cosmic rays (GCRs) are always present, although their intensity varies with solar activity. Solar cosmic rays are present only during very intense SPEs.

Scientists have recorded SPEs indirectly from ground observations since 1942 and directly from spacecraft since 1965. Between 1942 and 1953, the only way to detect SPEs was with ground-based instruments (ionization chambers and muon counters) designed to monitor the intensity of galactic cosmic rays. At that time only particles with extremely high energies (>4 GeV), high enough to penetrate Earth's magnetic shield to the top of the atmosphere, were detectable. As these cosmic rays passed through the atmosphere, they generated a nuclear cascade intense enough that some of the secondaries (primarily muons) reached the cosmic-ray recording instruments on the ground. When an SPE also brought such high-energy particles to Earth, the instruments recorded a transient rise in the count rate above the background set by galactic cosmic rays. From such signals, cosmic ray physicists estimated the dose rate (flux) and total dose (fluence) of the highest energy particles in the SPE. The early 1950s saw the development and deployment of cosmic-ray neutron monitors, which could observe particles with energies nearly an order of magnitude lower than those detected by muon counters or ionization chambers. The SPEs of solar cycle 19 (1954-1965), one of which (February 11, 1956) is famous for its high intensities at high energies, were recorded with neutron monitors.

Before 1957, cosmic ray physicists could infer the occurrence of large fluxes of lower energy particles (<100 MeV) only from the extra ionization they produced in the upper atmosphere. Such ionization absorbed radio signals, thus producing polar cap blackouts, which had (limited) use as a quantitative measure of particle flux or fluence. Since 1957, however, radio techniques have been developed that can measure the flux of lower energy particles through their ionospheric effects.[2] More important, the advent of the space age enabled direct observations of solar proton events at lower energies than can be monitored from the ground. Spacecraft studies of SPEs started in the early 1960s, with Explorer 12 in 1961, Explorer 14 in 1962, and Interplanetary Monitoring Program (IMP) 1 in 1963. Routine satellite measurements of SPEs, with good data coverage, started in 1965 and have continued almost uninterrupted over the intervening decades. At present, the NOAA geostationary operational environmental satellites (GOES) are responsible for supplying real-time measurements of energetic particles to SEC for SPE monitoring, while NOAA's National Geophysics Data Center archives SPE data for research. The resulting database, combined with data from heliospheric probes and from ground-based and balloon-borne instruments, has enabled studies that have greatly increased our familiarity with and understanding of SPEs.

Most research on SPEs has focused on characterizing individual events and, as data increased, on statistical studies of those characteristics. The current state of models for SPEs is assessed in Appendix A, in the context of their potential for contributing to the risk assessment effort. Although these studies have not yet revealed how particles are accelerated to relativistic energies or how to predict the flux or fluence from an individual solar event, they are useful for making statistically realistic estimates of the likelihood that SPEs might impact ISS construction.

2.2 PROBABILITY OF SPE IMPACT ON ISS CONSTRUCTION

Turner and Baker[3] first estimated the likelihood that SPEs will impact ISS construction using the significant-dose criterion (criterion 2, defined above). They statistically analyzed SPE data from the last four solar cycles to determine the average frequency of occurrence of class 2 SPEs (those capable of satisfying criterion 2) as a function of time from the solar minimum, then tabulated the intervals between the last solar minimum (near the beginning of 1997) and the times of the ISS construction flights, as specified in the then-current manifest, from June 1998 through June 2002. Assuming that the current solar cycle, cycle 23, will resemble the average of the previous four, they found that 11±6 class 2 SPEs would occur from June 1998 through June 2002. This means that between two and four class 2 SPEs probably would occur during an ISS construction flight. This result is easy to comprehend given that, according to Turner and Baker's statistics, the average probability that a class 2 SPE will occur during an arbitrary 2-week ISS construction mission is about 10 percent, and 34 shuttle missions are scheduled.

Turner and Baker's study raises an important question: Is it appropriate to apply statistics derived from averaging the previous four solar cycles (19 through 22) to the current solar cycle, considering the well-documented extent of intercycle variability? For example, the number of class 2 SPEs ranged from 10 to 30 over cycles 19 to 22. To address this question, NASA funded a workshop in late 1996 (near the end of cycle 22) to "assess [solar-cycle] prediction techniques and arrive at a reasoned consensus, including uncertainty, on how [cycle 23] will develop."[4] Because the record of sunspot numbers is longer than the record of SPEs, the workshop focused on predicting cycle 23's sunspot number history as an indirect measure of its potential for generating solar storms. It rated each technique on its relative ability to predict past cycles, then constructed a consensus forecast by taking an accuracy-weighted average over all techniques. The resulting consensus forecast predicted that cycle 23 should strongly resemble cycle 22 in amplitude and should peak in the spring of 2000. The panel revisited the question a year later, after the new cycle had begun, and confirmed the validity of this conclusion. Therefore, since cycle 22 produced stronger SPEs than the average of the last four cycles, 19 to 22, Turner and Baker's calculations probably underestimate the SPE impact on ISS construction flights.

To quantify for this report an exclusive cycle-22 perspective on the problem, Turner and Kemere[5] advanced cycle 22's SPE history forward 10 years to represent cycle 23. They used a cycle-22 SPE database compiled mainly from data archives at NOAA's National Geophysical Data Center (Margaret A. Shea, Air Force Research Laboratory, and Don F. Smart, Air Force Research Laboratory (ret.), personal communication). Turner and Kemere then superimposed the ISS-construction flight manifest (released in May 1998) on their ersatz history of cycle 23 SPEs and recorded the number of coincidences between class 2 SPEs and shuttle flights. To build a statistically robust result, they shifted the flight schedule forward by 2 weeks 47 times. For each iteration, they tallied the number of flights that would have experienced a criterion 2 SPE impact and expressed the result as a percentage of iterations per number of flights impacted. As shown graphically in Figure 2.1, 23 percent had one or two impacted flights, 47 percent had three or four impacted flights, 23 percent had five or six impacted flights, and 7 percent had seven or eight impacted flights. Note that all iterations had one or more impacted flights. These results quantify the expectation that the number of impacted flights is somewhat greater based on cycle 22 SPE history than on average SPE history.

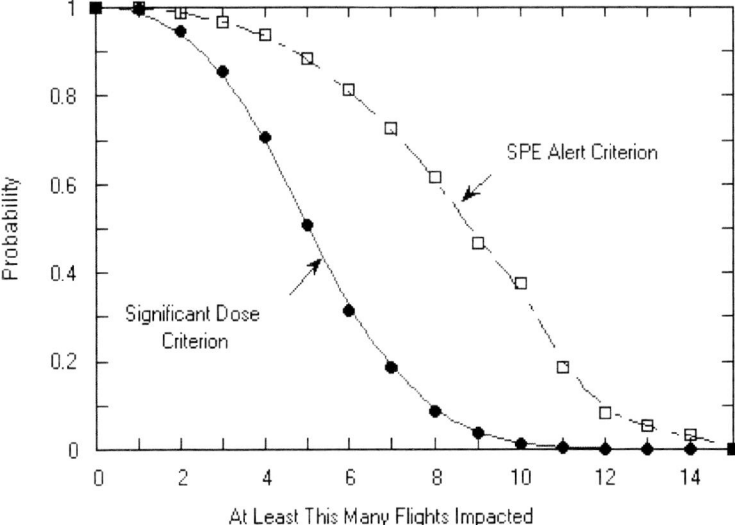

Figure 2.1 Cumulative probability curves for number of flights out of 43 that overlap an SPE according to the alert criterion (criterion 1) and the significant dose criterion (criterion 2).

For further verification of these predictions, Shea and Smart conducted an independent study similar to Turner and Kemere's except it used the SPE alert criterion (criterion 1) instead of the significant dose criterion (criterion 2) to identify flight-impacting SPEs. The criterion-1 study used the same (Shea and Smart) database as the Turner and Kemere criterion-2 study, projected the cycle 22 SPE history to derive an ersatz history for cycle 23, but assumed 10-day flights instead of 2-week flights. To build a statistically robust result, the hit count was iterated 48 times, each time advancing the flight schedule by 15 days. Figure 2.1 shows the results of the criterion 1 (Shea and Smart) and criterion 2 (Turner and Kemere) studies.

At face value, both curves in Figure 2.1 imply with near certainty that SPEs satisfying both criteria will impact at least two ISS construction flights. There is a 50 percent chance that significant-dose SPEs will impact five or more flights and that at least three additional flights will have an SPE alert. There is a 10 percent chance that significant-dose SPEs will impact eight or more flights and that at least four more flights will have an SPE alert. Although the exact number of impacted flights remains uncertain, it is inevitable that a flight director will have to respond to an SPE during a shuttle flight on more than one occasion, and probably on several occasions.

As is always the case with statistical predictions, there is a finite chance that the next solar cycle will produce considerably fewer SPEs than cycle 22. To make flight plan decisions based on this unrealistically sanguine hope, however, would be imprudent. In fact, if cycle 23 were not modeled on its predecessor, as was recommended by the solar physics community, the only reasonable alternative would be to evaluate a worst-case-cycle scenario, which would predict even higher risks of radiation. Radiation risk managers should plan, therefore, on significant-dose SPEs impacting more than one ISS construction flight.

2.3 CORRELATION BETWEEN SPEs AND SIZE OF THE SPE ZONE

A major factor in determining whether a given SPE poses a threat to astronauts in ISS orbit is the size of the SPE zones to which SPE particles have access, a topic discussed in Sections 1.1 and 1.3. The angular width of the SPE zones is normally around 30 degrees, which poses little if any threat to astronauts in ISS orbit. When the zones widen to the order of 45 degrees, as can happen during a major geomagnetic storm, 40 percent of the ISS orbit falls within the SPE zones and the radiation threat becomes nonnegligible if an SPE is in progress. Clearly, there is a tendency for the SPE zones to widen during SPEs. The tendency is not perfect: SPEs can run their course wholly ignored by the SPE zones, and SPE zones can open when there are no SPE particles to ingest. Nonetheless, as CSSP/CSTR demonstrates, the SPE zones often widen during high-dose SPEs, thus statistically favoring the worst-case combination. SPEs do not physically widen the SPE zones; rather, one aspect of a solar storm generates SPEs while another separate but correlated aspect widens the SPE zones.

The particles in an SPE can come from either a solar flare or the shock wave driven by a coronal mass ejection (CME), with the latter being responsible for the largest total doses. The shock-accelerated SPE particles travel toward and away from the Sun along the interplanetary magnetic field (IMF) threading the solar wind, but a significant fraction are trapped near the propagating shock by wave-particle interactions.[6] Thus, once the shock reaches Earth, the energetic proton flux can increase suddenly by as much as two orders of magnitude, making this shock spike the most dangerous portion of the solar particle event itself.

The SPE zones, on the other hand, are widened by the interaction between Earth's magnetosphere and the strong IMF associated with the CME-shock system. The spatial relations between the two sources of SPE particles and the regions of strong IMF accompanying a solar storm are illustrated in Figure 2.2. Consider, for simplicity, a situation in which Earth lies in the direct path of the oncoming CME and shock. As noted above, some SPE particles arrive at Earth long before the shock, while the IMF still exhibits its normal strength and the SPE zones occupy their normal volume. Consequently, preshock SPE particles generally have minimal access to the orbit of ISS, unless an earlier geomagnetic storm widened the zones and is still in progress (a scenario characteristic of solar maxima and so not unlikely during the ISS construction period). Conditions change after the shock passes, for SPE particles are still present and so, too, is a strong and often turbulent IMF, which in many if not most cases induces a magnetic storm. The SPE zones then widen and fill with the solar energetic particles in the ambient solar wind.

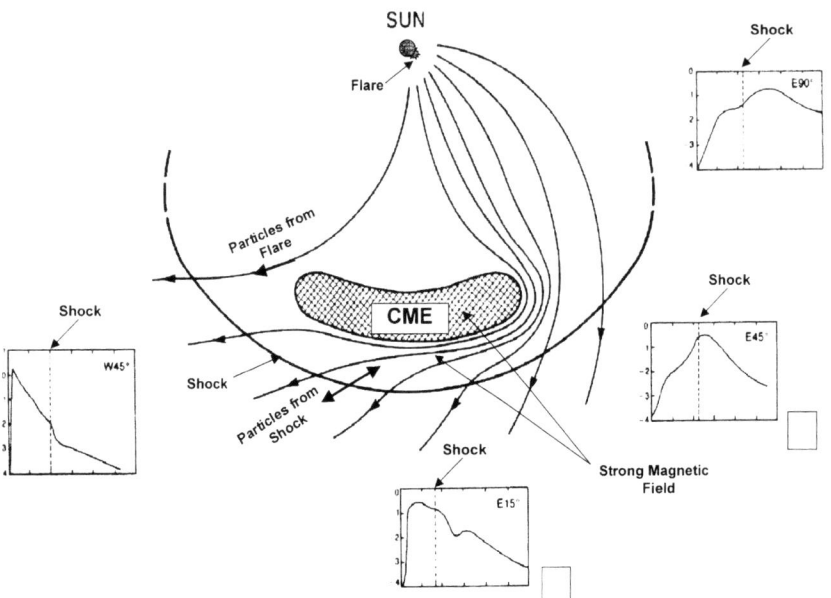

Figure 2.2 Flare, CME, shock, SPE particles, and strong IMF (after Cane et al.[7]). This figure illustrates the rise and fall of fluxes of solar energetic particles during an SPE. Starting from the Sun, we see the site where a flared occurred. The lines emanating from there are representative magnetic field lines, which the solar wind carries into interplanetary space. The lines have been bent by solar rotation, which gives them a clockwise curvature, and by the intrusion of a fast CME, which compresses them between its forward face and a shock wave that forms ahead of it. The energetic particles that make up the SPE are generated by the flare—these stream away from the Sun along magnetic field lines—and by the shock—these stream away from the shock in both directions, towards the Sun and away from it. The inserts show how the flux varies with time at different places relative to the site of the solar storm that caused the flare and the CME. The ticks on the vertical axes of the inserts mark off decades. The ticks on the horizontal axes mark off days. Insert A shows a direct hit. The Earth is centered in the path of the CME and receives high fluxes before and after the shock passes. Inserts B and C illustrate off-center hits, in which the site of the solar storm is to the east of the central meridian as seen from Earth. (This is astronomical east, which is defined by projecting Earth's direction of rotation onto the celestial sphere.) Insert D illustrates the case in which the site of the solar storm is west of the central meridian as seen from Earth. This case is magnetically connected to the flare site. Cases A, B, and C are dominated by particles generated at the shock. Case D is dominated by flare-generated particles.

From the perspective of radiation risk assessment, the double threat posed by high SPE fluence entering widened SPE zones is a flight director's nightmare. Aside from the amplifying effects of sequential solar storms, however, only a fraction of these events will combine the right elements to produce the worst-case scenario. These elements are primarily, though not exclusively, a fast CME launched near the central meridian of the Sun (the case discussed above), yielding the highest-dose SPEs and the biggest geomagnetic storms. Several situations can give rise to an SPE without a geomagnetic storm to widen the SPE zones: the IMF can point northward throughout the storm; the shock, CME, or both could miss Earth even though SPE particles reach it; or the SPE might come from a west-limb flare without an associated CME.

The tendency for major (Kp ~ 9) geomagnetic events to coincide with intense SPEs has been observed for individual storms, for example, those of July 1959, August 1972, October 1989, March 1991, and June 1991.[8] It is also evident from a correlation analysis between SPE intensity as measured by peak dose rate and geomagnetic disturbance intensity as measured by a standard magnetic activity index, Kp. The Kp index is for geomagnetism what the Richter scale is for seismology, a logarithmic proxy measure of the energy of disturbance. This 3-hour-averaged planetary index, which ranges from 0 (extremely quiet) to 9 (extremely disturbed), is a single-parameter measure of geomagnetic storm intensity that has been computed according to a uniformly applied algorithm continuously since 1932. It has become a standard against which space physicists like to correlate all manner of space weather variables. Geomagnetic conditions can be considered stormy when the Kp index exceeds 5, which happened 6.2 percent of the time during solar cycle 22. During class 1 SPEs (those capable of satisfying criterion 1 for an SPE event), however, Kp exceeded 5 nearly four times as often: that is, if an SPE is in progress, there is about a 24 percent chance that a geomagnetic storm is also occurring and that the SPE zones are dilated. For the most intense cases, the probability of a geoeffective storm coinciding with an SPE is still higher.

Since SPE zones widen during geomagnetic storms and Kp measures storm intensity, one can estimate SPE-zone width from observed Kp values. (There is, however, no database of SPE-zone widths that would allow us to eliminate the Kp proxy for them.) A Kp proxy for SPE-zone size can be obtained from an empirical relation that specifies in terms of Kp the angular size of a circle defined by the equatorward edge of the auroral oval, which is a rough indicator of the width of the SPE zones.[9] The equatorward edge of the auroral oval delimits the volume of the magnetosphere in direct plasma contact with the magnetotail, a volume into which energetic SPE particles can considerably (though not always completely) penetrate.

Figure 2.3 shows 3-hour averages of the directional dose rate from particles with energies above 30 MeV (which penetrate all parts of a space suit) recorded in 1989, a year around the maximum of solar cycle 22, during which several intense SPEs occurred. The figure shows that the most intense fluxes tended to occur when the SPE zones were wide, although the tendency was not always observed. For example, there were high fluxes (above 10^3 particles cm^{-2}s^{-1}sr^{-1}) when the width of the SPE zones was less than 30 degrees. These cases illustrate that the solar energetic particles get to Earth before the shock that brings the magnetic storm that opens the SPE zones. Nonetheless, a tendency for high fluxes to occur when SPE zones are wide is discernable. The box marks a danger

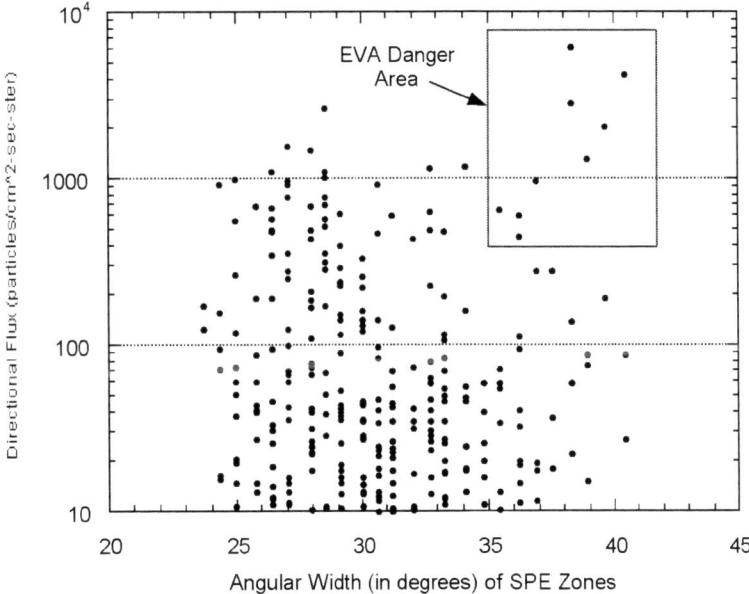

Figure 2.3 The SPE history for 1989 showing extreme SPE-zone dilation coinciding with extreme fluxes. Each point is a 3-hour average of the directional dose rate from particles with energies above 30 MeV. The statistical error is smaller than the dots. The angular width of the SPE zones has been estimated using Kp as a proxy. The box delineates the danger area for EVAs. (Courtesy of Don Smart.)

area where high dose rates combine with wide SPE zones to pose a hazard to astronauts in ISS orbit who are shielded only by a space suit. The nine points inside the danger area come from two storms during 1989. If ISS were inside the danger area while it was in the part of its diurnal cycle when it most deeply penetrates the SPE zones, it would spend a quarter to a third of each orbit exposed to SPE particles. During this time, an astronaut on a 6-hour EVA could receive a radiation dose between 10 and 100 percent of the short-term limits for eyes and skin.

2.4 SUMMARY AND RECOMMENDATION

CSSP/CSTR finds that a concentrated effort aimed at reducing SPE radiation risk to astronauts during ISS construction is needed. Based on the assumption—the best now available—that the radiation characteristics of the current solar cycle will resemble those of the last cycle, there is nearly a 100 percent chance that at least 2 out of 43 planned ISS construction flights will overlap a significant SPE and a 50 percent chance that at least 5 flights will overlap such an event. Moreover, whenever SPEs are in progress, the SPE zones show a marked tendency to widen over the polar latitudes reached by the ISS orbit, a tendency that becomes stronger with SPE severity. Two storms during 1989, near the maximum of the last solar cycle, illustrate the point. The SPE zones widened until they engulfed more than a quarter of the planned ISS orbit, while the radiation intensified enough to have pushed an astronaut over the short-term limit for skin and eyes during a single, ill-timed 6-hour EVA. CSSP/CSTR therefore recommends the rapid implementation of the following scientific elements of a program to reduce SPE radiation risk:

Recommendation 2: For real-time SPE risk management, carry out the steps needed to make usable by SEC and SRAG the models that use real-time data to specify the intensity of SPE particles and the geographical size and shape of the zones accessible to them.

This recommendation could be implemented early enough to have an impact on SPE radiation risk management during ISS construction; CSSP/CSTR views it as a high-priority item for action by NASA, NOAA, and the USAF. Chapter 4 and Appendix A of this report document other existing and potential resources that could contribute to implementing recommendation 2. Appendix A also addresses institutional issues related to developing the requisite modeling tools. Finally, chapter 5 includes a discussion of unresolved issues surrounding the "transitioning" of research models to operational use.

Recommendation 1 in Chapter 1 is relevant to Recommendation 2, set forth above. For these recommendations to succeed, flight directors must become involved in assessing the effectiveness of the new or improved tools for SPE risk management that could allow liberalizing the current (unofficial) flight rule on radiation that says changes in flight plans can be based only on real-time, on-site data.

2.5 NOTES AND REFERENCES

1. For fluence to dose conversion, see A.C. Tribble, *The Space Environment,* Princeton University Press, 1995, Figure 5.2.
2. D.K. Bailey, *Planet. Space Sci.*, 1964, p. 485.
3. R.E. Turner and J.E. Baker, "Solar particle events and the International Space Station," *Acta Astronautica*, 42, 1998, pp. 107-114.
4. J.A. Joselyn, J.B. Anderson, H. Coffey, K. Harvey, D. Hathaway, G. Heckman, E. Hildner, W. Mende, K. Schatten, R. Thompson, A.W.P. Thomson, and O.R. White, "Panel achieves consensus prediction of Solar Cycle 23," *EOS, Trans. Amer. Geophys. Union*, 78, 1997, pp. 205, 211-212.
5. R.E. Turner and C. Kemere, "Solar particle events and International Space Station," Report submitted to Committee on Solar and Space Physics. This material, which was presented to CSSP in August 1998, is available for viewing in the National Research Council's Public Access Records Office.
6. D.V. Reames, "Energetic particles and the structure of coronal mass ejections," in *Coronal Mass Ejections, Geophys. Monogr. Ser.*, 99, N.U. Crooker, J.A. Joselyn, and J. Feynman, eds., American Geophysical Union, Washington, D.C., 1997, pp. 217-226.
7. H.V. Cane, D.V. Reames, and T.T. von Rosenvinge, "The role of interplanetary shocks in the longitude distribution of solar energetic particles," *J. Geophys. Res.*, 93, 1998, pp. 9555-9567.

8. Z. Svestka and P. Simon, eds., *Catalog of Solar Particle Events, 1955-1969*, D. Reidel Publishing Co., 1975; H.E. Coffey, ed., *Collected Data Reports on August 1972 Solar-Terrestrial Events* (Parts 1, 2, and 3), World Data Center A for Solar-Terrestrial Physics, NOAA, Boulder, Colo., July 1973; M. Dryer, ed., *Space Science Reviews* (Special Issue), 19, 4/5, 1976; M.A. Shea and D.F. Smart, "Overview of the solar and interplanetary phenomena leading to the major geomagnetic disturbance on 24 March 1991, in *Workshop on the Earth's Trapped Particle Environment*, G.D. Reeves, ed., AIP Conference Proceedings 383, AIP Press, Woodbury, N.Y., 1996; and M.A. Shea and D.F. Smart, "Solar proton fluxes as a function of the observation location with respect to the parent solar activity," *Adv. Space Res.*, 17, 4/5, 1996, pp. 225-228.

9. M.S. Gussenhoven, D.A. Hardy, and M. Heinemann, "Systematics of the equatorward diffuse auroral boundary," *J. Geophys. Res.*, 88, 1983, pp. 5692-5708.

3

Relativistic Electrons and the International Space Station

3.1 OUTER BELT ELECTRONS

Solar particle events (SPEs), the subject of Chapter 2, are one of the two high-latitude radiation hazards to which astronauts on ISS construction crews can be exposed. The other is relativistic electrons—electrons with energies above 500 keV, which penetrate space suits—in the outer radiation belt. To assess the electron radiation hazard to ISS, CSSP/CSTR first clarified the extent to which the ISS orbit penetrates the volume of space where relativistic electrons occur and then considered how fluxes of relativistic electrons vary over time and on what they depend. It has thereby evolved empirical rules that can serve to trigger alerts and warnings of imminent increases in relativistic electrons and, after the increase occurs, to predict how the flux subsequently decreases. We begin with the relevant geometry.

The outer electron belt forms an Earth-circling torus, crescent-shaped in cross section, with the crescent concave earthward (see Figure 1.1). The apex of the heart of the outer belt extends roughly from 2.5 to 4 Earth radii (Re), although significant traces of the belt extend out to 10 Re. (1 Re is the unit used by radiation-belt physicists to represent distance; it equals 6,370 km.) Also, instead of referring to a crescent, they refer to L shells, where L is the distance (in Re) from the center of Earth to the apex of the crescent (or shell). Roughly speaking, an L shell is an Earth-circling surface generated by all magnetic field lines that cross the equator L Re from Earth's center. The crescent cross section of the radiation belts conforms to the dipolar geometry of the geomagnetic field.

L shells that define the outer radiation belt reach down to touch the atmosphere in each hemisphere in an annulus centered on the magnetic poles. In each hemisphere, the magnetic pole is offset from the geographic pole by 11.4 degrees. The heart of the annulus at the altitude of ISS extends from about 28 degrees to 31 degrees in angular width from its polar center. Since the geometrical considerations here are similar to those described in Section 1.3, we may use Figure 1.6, which gives the fraction of the ISS orbit exposed to the SPE zones for different zone sizes, to estimate the fraction of ISS orbit exposed to outer belt electrons. For this application, the appropriate zone size is about 33 degrees. (During highly relativistic events—discussed in Section 3.2—the outer radiation belt can extend down to 45 degrees, so by taking 33 degrees as a zone size, we are being conservative in the sense of not overestimating the seriousness of the situation.) Figure 1.6 shows that ISS spends between 0 and about 20 percent of its orbit in the relativistic electron annulus cycles every 24 hours. Although 20 percent seems like a small number compared with the 40 percent by which the SPE zones can engulf the ISS orbit during major geomagnetic storms, the accumulative effect can nonetheless be significant. Because ISS passes through the relativistic electron annulus every day, even during geomagnetic calms, as far as the geometry of the relativistic electron belt goes, the threat seldom goes away. Having established that during a part of nearly every day ISS spends about 20 percent of each orbit in the relativistic electron annulus, we look next at the dose rate during this time.

3.2 MONITORING OUTER BELT ELECTRONS

The flux of relativistic electrons in the outer radiation belt varies over time by many orders of magnitude. The variation is marked by events, called highly relativistic electron (HRE) events, that have a characteristic life cycle, rising quickly (on the order of 1 day) and decaying slowly (on the order of 5 to 10 days), although some events decay anomalously even more slowly. The fact that the onset of HRE events is well correlated with changes in solar wind conditions offers an opportunity to develop a protocol for HRE-event alerts and warnings that could demonstrate sufficient predictive value to be of operational use. The decay rate, which is faster on L shells closer to Earth,[1] seems to be stable enough to use to predict fluxes days ahead. To illustrate this point, we look at several HRE events.

Among the several satellites that currently record fluxes of relativistic electrons in the outer radiation belt are the Solar Anomalous Magnetospheric Particle Explorer (SAMPEX) and POLAR, two NASA research satellites. SAMPEX, which is in a low-altitude, 96-minute polar orbit, measures fluxes of electrons between 500 keV and 50 MeV in the low-altitude horns of the outer radiation belt. Plate 1 shows a color spectrogram from SAMPEX data of the average flux of electrons between 2 and 6 MeV as a function of L shell and day-of-year in 1997. During 1997, SAMPEX recorded 14 outer-belt, energetic-electron events during which the daily flux of energetic electrons jumped three orders of magnitude to values exceeding 10^4 electrons cm^{-2} s^{-1} sr^{-1}. The bottom panel of Plate 1 shows similar data from the POLAR satellite, whose highly elliptical orbit, ranging from 2 Re at perigee to 9 Re at apogee, covers L values from 2 to 100. The same events are seen in both data sets. Note that the integral flux of four of the events exceeded 10^5 electrons cm^{-2} s^{-1} sr^{-1}. Unfortunately for operational considerations, data from SAMPEX and POLAR satellites are not available in real time. Perhaps, however, a combination of LEO polar satellites that return data once per orbit and geostationary satellites that return data continuously in real time can assess electron flux in the outer belt completely enough to serve for critical operational purposes.

Two NOAA programs provide real-time satellite data on the space environment, including data on energetic electrons. One of these is the Polar-Orbiting Operational Environmental Satellites (POES) program, which collects data from satellites in low-altitude, polar orbits that cover the outer radiation belt, providing integral fluxes of electrons with energies >300 keV. POES data are taken at 840 km altitude, not far from the nominal 400 km altitude of ISS, and so are readily convertible to fluxes relevant to EVAs at ISS. The conversion must reduce (usually by a factor of less than 2) the >300 keV fluxes that POES provides to fluxes of >500 keV electrons relevant to EVAs, and it must compensate for the different altitudes of the two orbits (fluxes at ISS are about two-thirds of those at POES). The fluxes of >500 keV electrons at ISS are thus about 30 percent of >300 keV electron fluxes that POES measures on the same L shell.

Figure 3.1 illustrates POES energetic electron measurements. The first and third panels show 12-hour running means of fluxes in the outer belt (L from 4.5 to 4.8 Re) throughout 1997. The maximum average flux was often above 10^4 electrons cm^{-2} s^{-1} sr^{-1}, and on one pass it exceeded 10^5 electrons cm^{-2} s^{-1} sr^{-1}. It can be seen that the storm enhancements in Figure 3.1 decay at a rate that is about the same from event to event, forming a basis for the prediction capability described in Section 3.3.

The second NOAA program that provides real-time satellite data on outer belt energetic electrons is GOES. Satellites in this program also measure fluxes of SPE ions, solar X-ray flux, and the magnetic field at L = 6.6 (geosynchronous orbit). Daily averages of >2 MeV electrons recorded on GOES are shown in the second and fourth panels of Figure 3.1. Since L = 6.6 is beyond the core of the outer radiation belt, GOES fluxes are generally less than the core values that POES record. In 1997, the maximum daily value from GOES was about 10^3 electrons cm^{-2} s^{-1} sr^{-1}, compared with 10^5 electrons cm^{-2} s^{-1} sr^{-1} from POES. Nonetheless, it can be seen that GOES fluxes track POES outer belt fluxes, which means that GOES can serve as a continuous proxy monitor of the intensity of outer belt electrons.

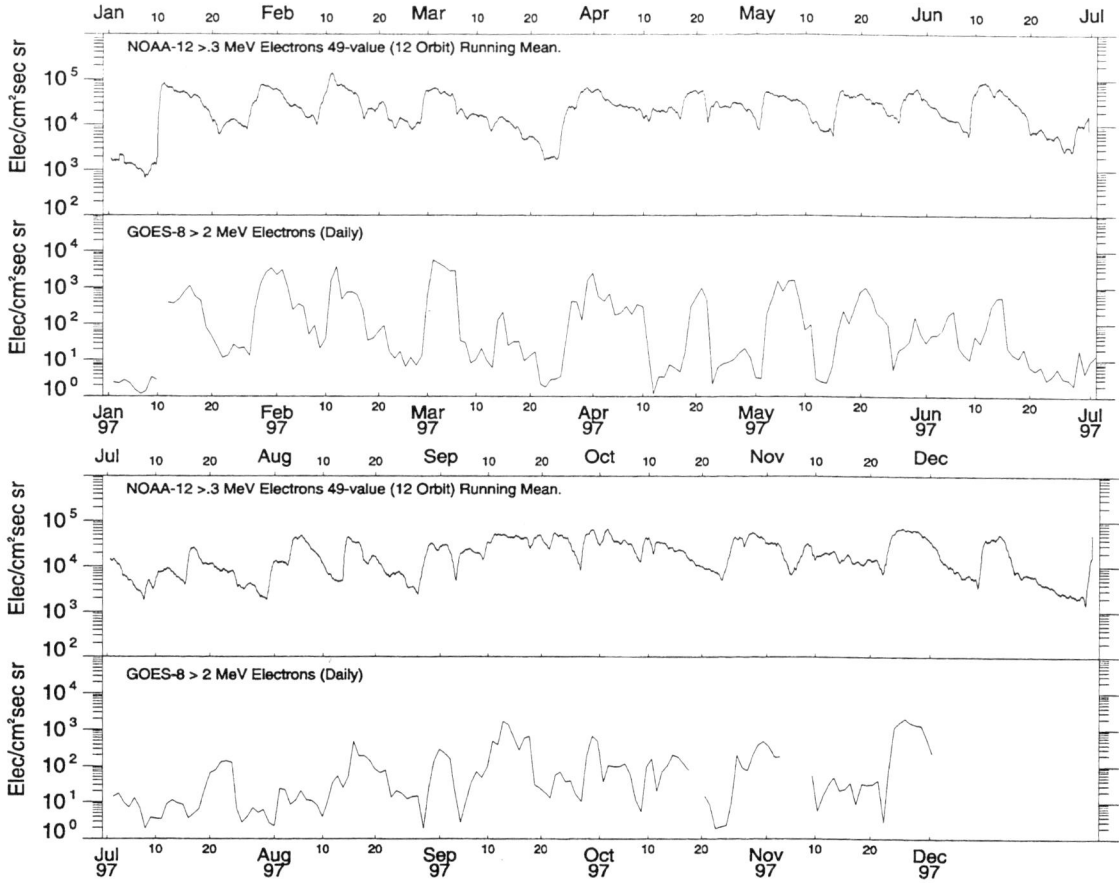

Figure 3.1 Average energetic electron fluxes recorded on POES (NOAA-12) and GOES: January to June 1997 (top half) and July to December 1997 (bottom half).

3.3 PREDICTABILITY OF RADIATION BELT ELECTRONS AT LOW ALTITUDE

The HRE events of 1997 shown in Figure 3.2 fall into two categories: those with a sharp rise and exponential decay to background levels over several days and those that stay elevated for many days. These two categories also pertain to HRE events recorded by other satellites, such as OGO 5,[2] DMSP,[3] POLAR,[4] and SAMPEX.[5] Decay rates of several days agree with theoretical calculations using diffusion and atmospheric loss processes. Longer decay rates have been attributed to an additional acceleration of ambient plasma around $L = 6$.[6]

The rapid rise and exponential decay events recorded by NOAA-12, a POES, permit developing and testing a prediction model that can be applied to such events. Figure 3.2 gives NOAA-12 data for the January 1997 event seen in Figure 3.1. The points in the top panel show the intensity of radiation in the core of the outer belt. The points are fluxes averaged between $L = 4.5$ and 4.8 (the core of the outer belt) for each orbit during the event. To smooth the points so that an exponential curve can be fitted to the decay phase, the second panel gives a running mean over 49 passes through the outer belt, which covers about 12 hours and 13 orbits. (NOAA-12 typically passes though the outer belt four times during every orbit.) The bottom panel, which shows >2 MeV electron fluxes recorded by GOES-8, illustrates the ability of a geostationary satellite to continuously monitor the electron intensity during such an event.

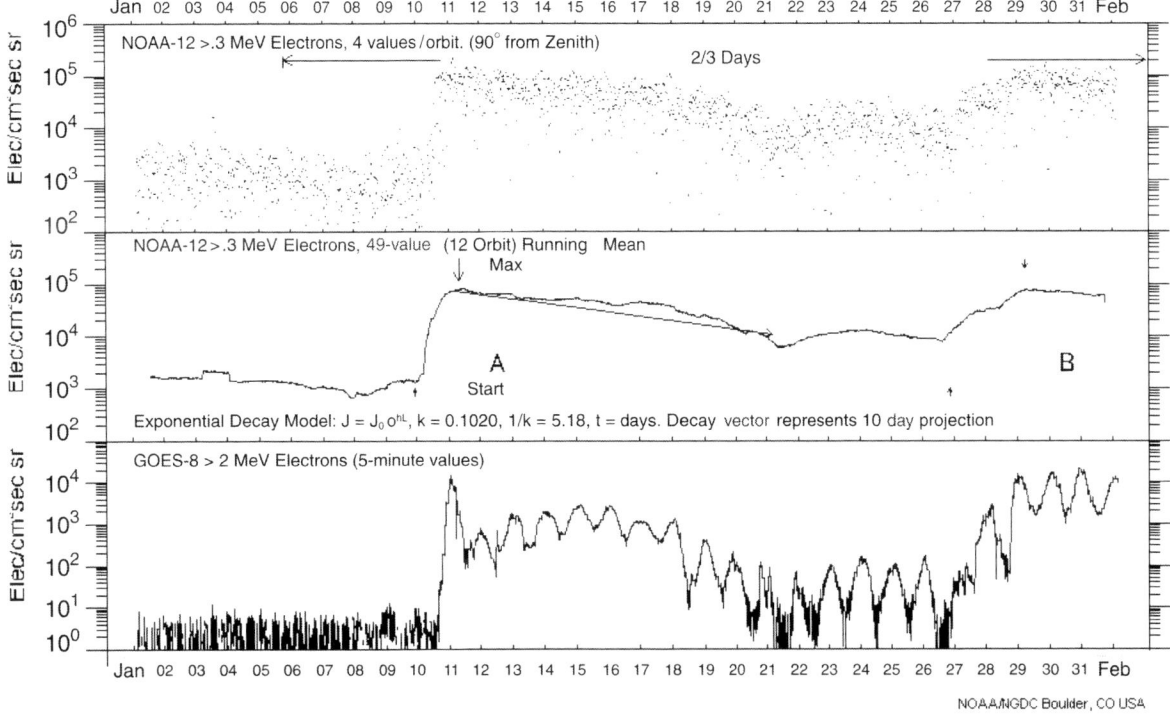

Figure 3.2 Energetic electron fluxes recorded by NOAA-12 and GOES-8 during an HRE event in January 1997.

In the second panel of Figure 3.2, the line through the decay phase of the event is an empirical fit of the form $J = J_o e^{t/\tau}$, where J is the integral electron flux, t is time in days, and τ (= 5.184) is the decay time in days. Others have reported a similar decay rate from analyses of similar data recorded on several different satellites.[7] With the decay time set at 5.18 days, this procedure was applied to nine other rapid rise and exponential decay events captured in the NOAA-12 data for 1997. The results are shown in Table 3.1, with event A being the event taken here as the standard, the January event of Figure 3.2. The columns of numbers under each event give the percentages by which the measurements differ in absolute value from the predicted flux based on the standard exponential fit. As a rule, one can predict the average flux of energetic electrons up to 6 days after the onset of a major enhancement (day 1) before the percent deviation exceeds 100 percent. For particle fluxes that vary by four orders of magnitude, this degree of predictability is rather remarkable. It could probably be improved by taking account of the observed decrease in τ for smaller L shells.[8]

3.4 ASSESSMENT OF HAZARDS FACED BY ASTRONAUTS DURING ISS CONSTRUCTION

To construct a worst-case scenario for assessing the risk astronauts face of being exposed to HRE events, CSSP/CSTR considered event C in Table 3.1, the extreme event of 1997, which was not an active year. The maximum running-average flux recorded between L = 4.5 and 4.8 was 1.3×10^5 electrons cm^{-2} s^{-1} sr^{-1}. If an EVA of maximum duration (6 hours) were to coincide with the maximum duration (6 hours) of this event, the fluence of >300 keV electrons encountered by an astronaut in the NOAA-12 orbit would have been 1.5×10^9 electrons cm^{-2}. Since the ISS inclination is less than the NOAA-12 inclination, the ISS spends about twice as long per orbit in the outer radiation belt and would experience a correspondingly larger fluence. Taking all factors into account, including the 30 percent reduction of NOAA-12 fluxes discussed earlier, the fluence from relativistic electrons

Table 3.1 Absolute Difference Between Model Prediction and Smoothed Flux (in percent)

Day	Event										Aver.
	A	B	C	D	E	F	G	H	I	J	
1	0	0	0	0	0	0	0	0	0	0	0
2	3	13	31	22	6	7	9	11	12	14	13
3	3	18	17	33	23	28	34	39	45	50	29
4	12	28	15	21	8	6	5	3	2	0	10
5	29	21	14	12	10	18	27	36	44	53	26
6	25	2	14	17	2	6	10	14	18	22	13
7	45	13	156	3	103	134	166	197	228	259	130

during an ill-timed EVA at the same time as this HRE event would have been about 10^9 electrons cm^{-2}, which corresponds to a dose of about half a sievert[9] (see Section 1.3 for definitions of radiation units). Several events during 1998 reached four times the dose rate of this 1997 event (J.B. Blake, Space Sciences Department, Aerospace Corp., personal communication, 1999). Although the shielding provided by a space suit will reduce this number by a factor of about two, the dose would nonetheless be great enough to force an astronaut over the short-term limit for skin and eyes.

Another threat posed by the high-energy electron environment is the electrical charging of solid objects in space, such as pieces of the ISS and the astronauts' space suits. This is probably not a health hazard, but it might result in an electrostatic discharge when an astronaut touches something. If that something is connected to an electrical circuit, the circuit might be damaged. Electrostatic discharges in the outer electron belt are a familiar source of problems for spacecraft circuits. Although such a threat falls outside the purview of this report, it is raised to illustrate the need for communication between NASA centers that deal with different aspects of the radiation problem. This is the subject of Chapter 6.

3.5 OPERATIONAL STRATEGY

A strategy for developing an operational capability to predict high-energy electron events of concern to astronauts on the ISS should include the following elements:

• *Real-Time Event Identification from GOES Data.* An operational system to predict the flux of energetic electrons in the outer radiation belt using the scheme described above must monitor particle fluxes in a timely manner. The bottom panel of Figure 3.2 shows that the enhancements in GOES-8 electrons occur simultaneously with the enhancements recorded on NOAA-12. GOES data arrive at the Space Weather Operations Center (SWOC) of SEC within a few seconds of measurement. By contrast, NOAA-12, which spends only about 10 percent of its time in the outer belt, transmits its data to ground only once an orbit, or every 101.5 minutes.

• *Routine Monitoring of Low-Altitude, Outer-Belt Electrons.* Electrons exceeding 300 keV in the outer radiation belt recorded by the Space Environment Monitor on all POES should be routinely averaged and a running mean prepared, from which a regular 12-hour forecast can be made. This procedure would allow predictions for each L shell based on a decay rate appropriate to the shell.

• *The Forecast Model.* In general, until a realistic dynamic model can be developed, the following approach should yield useful results. If no event is in progress, persistence should dominate the forecast. If a rapid onset and exponential decay event resulting from reasonable pitch angle diffusion is in progress, then the equation described in Section 3.3 ($J = J_o e^{t/\tau}$) is a better choice than persistence. If a gradual decay event is in progress, then persistence again is the best option until a realistic model can be developed. Since the decay time is slow, the data from POES every 101.5 minutes are adequate to monitor progress of the decay of radiation fluxes to see if the prediction based on exponential decay is holding and to update the prediction.

The strategy described here applies to predicting the flux of relativistic electrons as it decays after the sudden rise (on the order of a day) that characterizes the onset of an HRE event. As mentioned earlier, the onset of HRE events is correlated with changes in solar wind conditions. A sudden, significant increase in solar wind speed (for example, more than 100 km/s in less than 1 day) and southward-pointing IMF are relatively good predictors of the onset of an HRE event.[10] Such sudden, significant increases in solar wind speed are common at the leading fronts of fast solar wind streams, which might or might not be accompanied by a southward-pointing IMF. An upstream solar wind monitor can provide the input needed to implement a protocol for issuing alerts and warnings based on this correlation.

3.6 SUMMARY AND RECOMMENDATIONS

Nearly every day an ISS orbit transects the outer radiation belt, where relativistic electrons reside. When the length of transection is greatest, ISS spends about 20 percent of its time in the belt. During relativistic electron events, which happen on average about once per month, the intensity of relativistic electrons in the belt increases by as much as four orders of magnitude. During an ill-timed EVA, when the intensity of relativistic electrons is greatest, the dose received by an astronaut could be several tenths of a sievert, which could put him or her over a radiation dose limit. Procedures can be implemented to specify and forecast at least approximately the intensity of relativistic electrons in the outer belt. POES provide measurements of relativistic electron fluxes that are transferable to the ISS environment with reasonable accuracy. These measurements are available about every hour and a half. GOES provide relativistic electron measurements continuously, but they are not as directly transferable to the ISS orbit. They do, however, track POES measurements. Thus, in combination, POES and GOES measurements allow radiation risk managers to follow the variation of electron intensity in the outer belt. A crucial piece of hardware that the ISS project should provide is an electron dosimeter outside the station. This would allow SRAG to assess the correctness of the specifications and forecasts based on measurements by POES and GOES.

The spatial (as well as temporal) variability of energetic electrons must also be taken into account before the total dose on a particular EVA schedule can be predicted. To make accurate predictions, several other things must be done, including the following: (1) determining the decay time of HRE events as a function of the L shell; (2) converting >300 keV electron flux, which is measured, to >500 keV electron flux, which is desired (a rough conversion factor of 30 percent was discussed); (3) estimating the size of the "loss cone," which is needed to compute the flux at the position of ISS (latitude, longitude, and altitude) from the position of GOES or POES (to allow improving the 30 percent estimate and continuously updating it); (4) distinguishing between events with rapid rise and exponential decays and events with a more gradual decay; and (5) constructing a reasonable dynamic model of the radiation belts with physics included.

This chapter has cited evidence for a good correlation between changes in solar wind conditions and the onset of HRE events. A project to develop an HRE-event-onset protocol has a reasonable chance of producing a product reliable enough to be of operational use in issuing alerts and warnings.

Recommendation 3a: NASA should implement a procedure for using POES and GOES measurements of relativistic electrons in the outer radiation belt to specify and forecast the electron radiation environment at ISS. (Such a procedure is outlined in Section 3.3.)

Recommendation 3b: As soon as possible, JSC should install an electron dosimeter and an ion dosimeter outside the ISS that can return data in real time to SRAG at JSC.

Recommendation 3c: A project should be initiated to develop a protocol for identifying the conditions that produce highly relativistic electron events based on the demonstrated good correlation between changes in solar wind conditions and the onset of such events. The recommended project might be a candidate for support by one of the affiliated agencies of the National Space Weather Program (NSWP). (See Section A.5.)

3.7 NOTES AND REFERENCES

1. M. Schulz and L.A. Lanzerotti, *Particle Diffusion in the Radiation Belts,* Springer-Verlag, 1974; D.N. Baker, J.B. Blake, L.B. Callis, J.R. Cummings, D. Hoverstadt, S. Kanekal, B. Klecker, R.A. Mewaldt, and R.D. Zwickl, "Relativistic electron acceleration and decay time scales in the inner and outer radiation belts: SAMPEX," *Geophys. Res. Lett.,* 21, 1994, pp. 409-412.

2. H.I. West, Jr., R.M. Buck, and G.T. Davidson, "The dynamics of energetic electrons in the Earth's outer radiation belt during 1968 as observed by the Lawrence Livermore National Laboratory's spectrometer on OGO 5," *J. Geophys. Res.,* 86, 1981, pp. 2111-2142.

3. M.S. Gussenhoven, E.G. Mullen, R.C. Filz, D.H. Brautigam, and F.A. Hanser, "New low-altitude dose measurements," *IEEE Trans. Nuc. Sci.,* NS-34, 1987, p. 676.

4. J.B. Blake, W.A. Kolasinski, R.W. Fillens, and E.G. Mullen, "Injection of electrons and protons with energies of tens of MeV into L < 3 on 24 March 1991," *Geophys. Res. Lett.,* 19, 1992, p. 821.

5. D.N. Baker, J.B. Blake, L.B. Callis, J.R. Cummings, D. Hoverstadt, S. Kanekal, B. Klecker, R.A. Mewaldt, and R.D. Zwickl, "Relativistic electron acceleration and decay time scales in the inner and outer radiation belts: SAMPEX," *Geophys. Res. Lett.,* 21, 1994, pp. 409-412.

6. R.S. Selesnick, J.B. Blake, W.A. Kolasinski, and T.A. Fritz, "A quiescent state for 3 to 8 MeV radiation belt electrons," *Geophys. Res. Lett.,* 24, 12, 1997, pp. 1343-1346.

7. West et al., 1981; Gussenhoven et al., 1991; Blake et al., 1992; and Baker et al., 1994.

8. See note 1.

9. For fluence to dose conversion, see A.C. Tribble, *The Space Environment,* Princeton University Press, 1995, Figure 5.2.

10. J.B. Blake, D.N. Baker, N. Turner. K.W. Ogilvie, and R.P. Lepping, "Correlations of changes in the outer-zone relativistic-electron population with upstream solar wind and magnetic field measurements," *Geophys. Res. Lett.,* 24, 1997, pp. 927-929. This reference contains references to earlier work on the subject.

4

Spacecraft Sources of Operational Radiation Data

4.1 VALUE OF SPACECRAFT MONITORS IN SUPPORT OF ISS CONSTRUCTION

NASA operates a number of spacecraft able to provide space weather information relevant to managing radiation risk during the construction and operation of the International Space Station (ISS). Particularly useful in this regard are spacecraft at the upstream Lagrangian point (L1), the position where the Sun's gravity and Earth's gravity balance to form, curiously enough, a virtual planet around which satellites can orbit in a plane perpendicular to a line connecting the Sun and Earth. This orbital plane is about a half-hour to an hour of solar-wind travel time upwind from Earth, depending on solar wind speed. Consequently, these spacecraft can warn of an impending increase in the radiation level at ISS when a shock impacts the magnetosphere. By continuously measuring the solar wind and the IMF, they detect a CME's bow shock a half hour or more before it reaches Earth and after that the CME itself, and so give data from which to predict whether and by how much the SPE zones will widen. Moreover, they continuously monitor the full, unshielded intensity of the solar energetic particles, which specifies the radiation intensity within the SPE zones.

As an example of the radiation profiles that are relevant, Figure 4.1 shows a series of severe SPEs that included the most intense event of solar cycle 22 (that event produced some of the points in the danger area delineated in Figure 2.3). Arrows in Figure 4.1 point to three sudden rises in intensity at the Earth, with the first one being the most intense. These sudden rises were associated with flare explosions on the Sun that released high intensities of X rays; the first flare occurred near the center of the solar disk. Following the initial rise after the first event, there was a second increase that was due to the passage of an interplanetary shock. The shock passage occurred somewhat more than 1 day after the initial flare and was associated with the highest intensities of the entire period. The second and third events in the series did not have similar shock-associated increases in flux.

The data in Figure 4.1 show the essential features of SPEs and illuminate the extent to which L1 spacecraft can warn of increases in radiation level in the polar regions of interest to ISS. Although active regions on the Sun that produce SPEs like the one shown in Figure 4.1 can be observed for a long period (up to months in some cases), until recently the ability to predict when an actual region might erupt into a large flare has been poor. Although a propensity to flare could be noted, each flare onset was essentially a surprise. Recently, however, a characteristic sigmoidal structure in X-ray images has been identified that may be a reliable precursor of CMEs.[1] This precursor should allow forecasters to predict the location of a CME (the location of the sigmoid). However, forecasters would not know precisely when, within a period of a few days, the CME would occur.

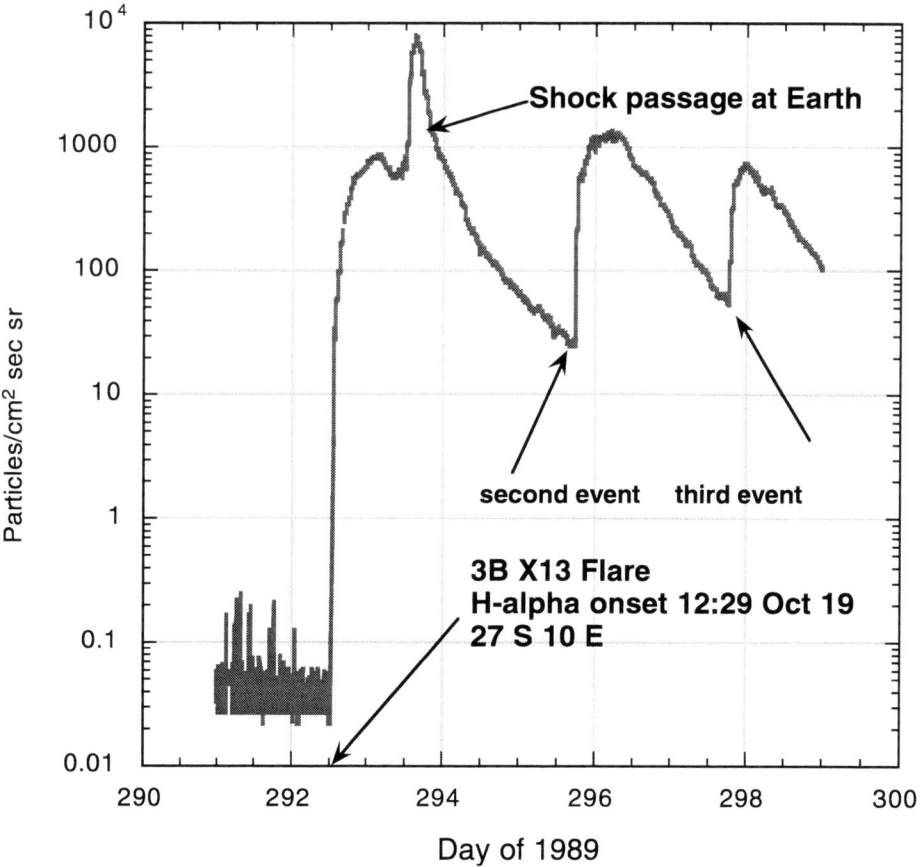

Figure 4.1 Intensities of protons near Earth during the most intense series of SPEs in solar cycle 22. Particles with the energies shown in this figure (>30 MeV) can penetrate spacecraft hulls.

As mentioned in Section 2.3, solar flares and CMEs contribute energetic particles to SPEs, although the particles they produce have different time profiles, compositions, fluences, and distributions in space. Compared with CME particles, flare particles generally arrive at Earth more promptly, have higher fluxes of relativistic electrons, are confined to a fairly narrow region of space that is magnetically connected to the flare site (on average, Earth is magnetically connected to flare sites at about 60 degrees west solar longitude), and have smaller total fluence. Compared with flare particles, CME particles generally take longer to arrive at Earth, have fewer relativistic electrons, reach higher energies, are spread over a wide area centered on the central meridian of the Sun, and have greater total fluence. Both CMEs and flares can produce their characteristic kind of SPEs in the absence of the other.[2]

A CME is an injection of solar material into interplanetary space at speeds often in excess of 1,000 km/s (2,000,000 mph). A typical travel time to Earth is about 48 hours. (The CME arrival in Figure 4.1 was atypically fast.) If a CME moves fast enough (supersonically with respect to the solar wind), a shock wave forms ahead of it that energizes particles as it moves through the interplanetary medium. According to one model of shock acceleration, a shock can produce both a long SPE profile, lasting 1 or 2 days, and a shock spike such as that seen in the first SPE in Figure 4.1.[3] The severity of a shock-generated SPE depends on the speed of the parent CME and on its launch site on the Sun; the most intense cases are associated with events that occur near the center of the visible disk, as was the case for the event shown in Figure 4.1.

4.2 AN INTERAGENCY FLEET OF SPACECRAFT MONITORS

The data on the October 1989 event shown in Figure 4.1 were taken by instruments on the GOES series of spacecraft, which are in geostationary orbit, not at L1. Since then, the International Solar-Terrestrial Physics (ISTP) program has launched a suite of spacecraft, now at L1 and in Earth orbit, that, taken together with others such as Advanced Composition Explorer (ACE), SAMPEX, and Yohkoh, provide an unprecedented opportunity to attempt to predict and monitor any large SPE (see Table 4.1). Plate 2 illustrates this for the January 6-11, 1997, event, during which the Solar and Heliospheric Observatory (SOHO) observed the launching of a CME from the Sun on January 6. As often happens, this CME produced no SPE. The arrival time of the CME was predicted, and on January 10-11, the prediction was confirmed when a very powerful shock passed by the SOHO and WIND spacecraft, both of which were at L1. Thus, SOHO and WIND were able to directly detect the shock about half an hour before it impacted Earth. Currently, upstream monitoring is carried out using real-time data from the ACE spacecraft; a sample of ACE data taken from the Web site is shown in Figure 4.2. Plate 3 shows the correlation between solar wind speed increases observed during this period and radiation belt intensities as measured at about 600 km, somewhat higher than the ISS orbit. Although there were no SPE intensity increases in this case, the impact of the CME on Earth's magnetosphere greatly increased the radiation levels, as shown in the right-hand panel.

Since Earth's magnetic field shields ISS from solar particle radiation during most of its orbital path, the actual exposure will depend on the orbital location of ISS when radiation levels peak. Figure 4.3 (upper panel) illustrates the geometry of interplanetary particle access to ISS. In the panel, a series of ISS orbital tracks is superposed on a globe. The gray area in the polar regions marks the geographic location of energetic ions observed on SAMPEX during a large SPE (in November 1992). The gray area is offset from the ISS orbit because Earth's magnetic field is offset from the spin axis, and it can be seen that this causes some ISS orbits to pass into the gray zone of particle access, while other orbits miss it entirely. This effect was described quantitatively in Figure 1.5. Since it takes 24 hours for the ISS orbital track to repeat itself over the globe, a warning can be given to ISS many hours in advance if its location above Earth will not take it through the access region for a number of orbits. Of course, if the timing is unfavorable, there might be little or no additional warning time before ISS enters the polar region beyond the hour or so offered by ACE and WIND.

Figure 4.3 (bottom panel) illustrates another complication of ISS penetration into the region of polar cap access, namely that the passage of the CME perturbs the Earth's magnetic field, and this perturbation increases the size of the polar access region for hours at a time. Figure 4.3 shows the access region's boundary as measured on SAMPEX, along with a proxy location determined by a ground-based magnetic activity index (Dst). Note that the directly measured boundary moved before passage of the shock and before the change in the Dst index. Although this case is unusual in that the SPE zone widened before the storm started, it nonetheless suggests that the surest way to accurately determine the location and size of the SPE zones during a large shock event is direct particle measurement by satellites in low Earth polar orbits that cross the boundaries four times per orbit (95 minutes).

Table 4.1 Spacecraft and Data Sources Relevant to ISS Construction and Operation

Spacecraft	Location	Data Available	Real-Time Data?
SOHO	L1	CME early warning	20 h/day
Yohkoh	Near Earth	X-ray flare images	Once per day
ACE	L1	Solar wind, field, and particle data	~24 h/day
WIND	Nearer Earth	Solar wind, field, and particle data	Flexible
GOES	Geosynchronous orbit	Particle data and X-ray flare image	24 h/day
POES	Near Earth	Particle data, including cutoffs	24 h/day
DMSP	Near Earth	Particle data, including cutoffs	24 h/day
SAMPEX	Low Earth orbit, high inclination	Particle data, including cutoffs	Twice per day

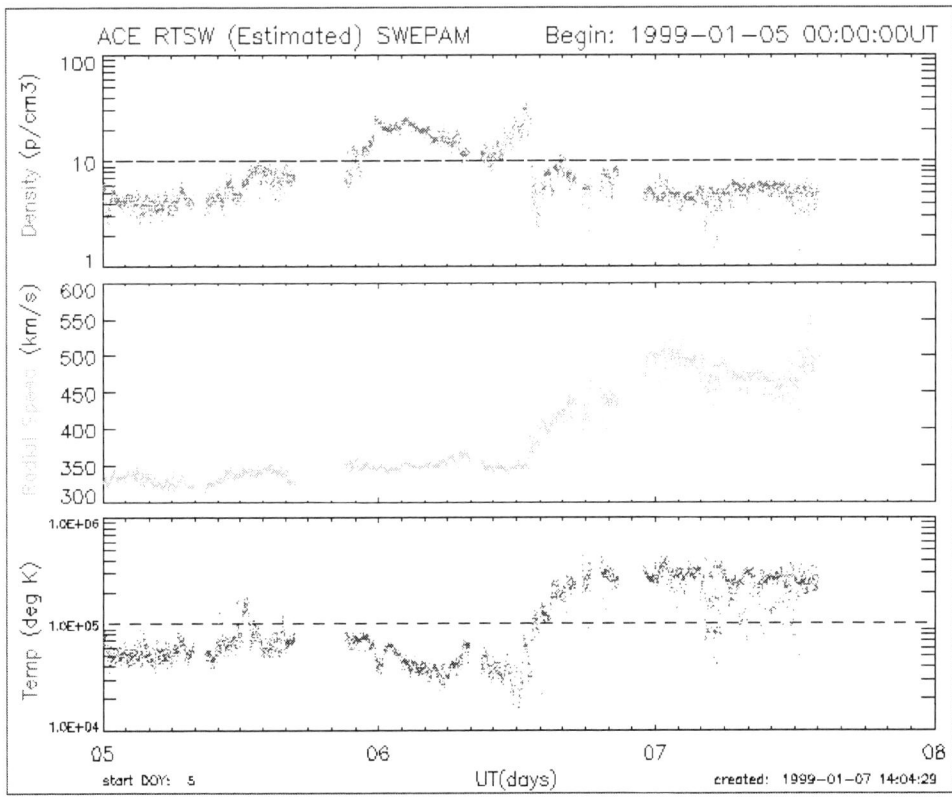

Figure 4.2 An example of near-real-time data available from the ACE spacecraft. The plot shows data obtained from January 5 to 8, 1999. From top to bottom, the quantities plotted are solar wind density (particles/cm^3), solar wind speed (km/s), and solar wind temperature (K). In addition to this information on solar wind, information on the magnetic field and energetic particle intensities is also available. All the real-time data from ACE are processed on board for conversion to scientific units and then telemetered continuously to ground stations worldwide. The data are collected by NOAA's Space Environment Center and made available on the Web.

Table 4.1 lists NASA Sun-Earth Connection, NOAA, and Department of Defense (DOD) spacecraft that provide relevant measurements and their capabilities for delivering data in the near-real-time manner that would be needed if required for ISS construction or operations. ACE, SOHO, GOES, the Defense Meteorology Satellite Program (DMSP), and POES gather a large quantity of near-real-time data; WIND and SAMPEX are more limited, but the amount of coverage could be increased if necessary. Other missions, such as Yohkoh, that indirectly sense the liftoff of CMEs through the brightening in X rays of magnetic arcades could also play a role in identifying candidate sources of SPEs.

4.3 FUTURE SPACECRAFT IN SUPPORT OF ISS OPERATIONS

Several future missions are expected to advance the study and forecasting of space weather. The High Energy Solar Spectroscopic Imager (HESSI) will be launched in mid-2000 to explore the basic physics of particle acceleration and energy release in solar flares by carrying out simultaneous, high-resolution imaging and spectroscopy of solar flares, from 3 keV X rays to 20 MeV gamma rays, with high time resolution. Like Yohkoh, HESSI

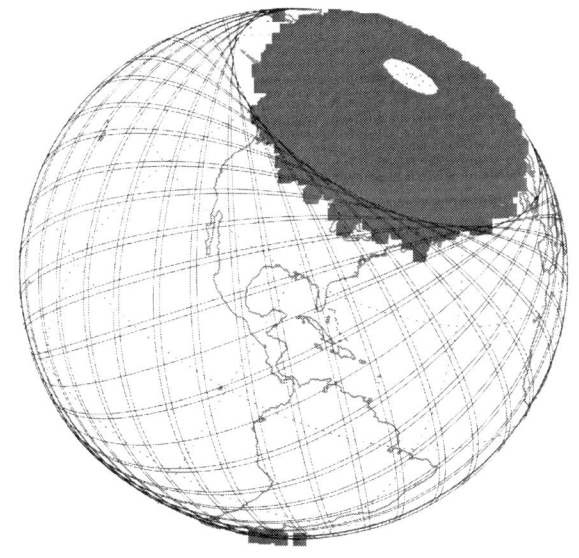

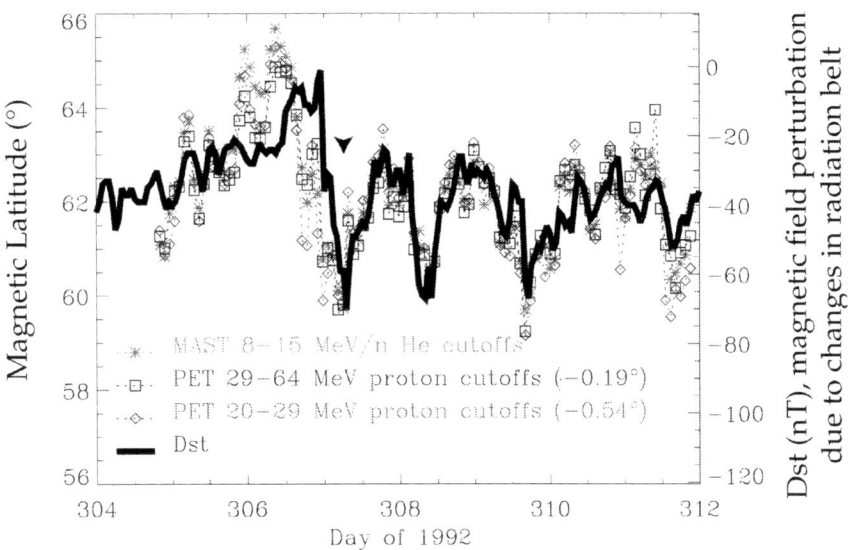

Figure 4.3 Top panel: Location of ground track of ISS orbits superposed on a globe along with polar cap areas (shown in gray), where multimega-electron volt solar energetic particles penetrated to low altitudes during an SPE in November 1997. (Courtesy of Leske et al.[4]) Bottom panel: time variation of the lower boundary of the polar cap measured by SAMPEX during the November 1997 SPE. The arrow marks the time of arrival of an interplanetary shock. The particle cutoff latitude measured on SAMPEX moved from ~66 degrees to ~60 degrees many hours before the arrival of the shock and even before Earth's magnetic field (heavy black line) registered a change.

can give flare warnings. A systematic study of CMEs and their development and effects near Earth is planned for the Solar Terrestrial Relations Observatory (STEREO) mission; however, its projected launch date of 2004 is later than the current schedule for ISS construction. The Inner Magnetospheric Explorer (IMEX) is scheduled for launch in mid-2001, just after the peak of solar cycle 23. Its mission is to study the dynamics of the inner magnetosphere during major geomagnetic storms. It will be the first mission in the inner magnetosphere to contain a full complement of field and particle detectors (especially electric fields) while there is a full-time upstream monitor in the solar wind (the ACE mission). In addition to these NASA missions there will be NOAA/DOD missions in the GOES/POES series that will include a Solar X-ray Imager (SXI).

4.4 SUMMARY AND RECOMMENDATION

A strategically placed fleet of spacecraft is currently taking data that can provide information on the radiation environment of the ISS orbit in real time and in advance of real time. Spacecraft in geostationary and L1 orbits monitor the Sun and its corona in multiple wavelengths and so can diagnose flare potency and warn of oncoming CMEs with considerable skill (see Section A.4). They also monitor SPE fluxes to give direct information on the radiation intensity within SPE zones. L1 spacecraft monitor solar wind and IMF parameters that can be used to predict the size and shape of SPE zones. Spacecraft in relatively low-altitude polar orbits monitor the flux of relativistic electrons in the outer radiation belt, which the ISS orbit transects. The information will provide flight managers with real-time, high-quality radiation-risk parameters. What is needed is a mechanism to channel the relevant information to the Space Radiation Analysis Group at Johnson Space Center.

Recommendation 4: Promptly convene a meeting of pertinent NASA Space Science Enterprise, SRAG, and SEC managers with the principal investigators of satellite instruments. The meeting would (1) consider ways to extend the capabilities of the current spacecraft fleet to provide real-time radiation data for driving models and specifying the ISS radiation environment and (2) formulate an implementation plan for swiftly channeling such data to radiation risk managers at JSC.

4.5 NOTES AND REFERENCES

1. R.C. Canfield, H.S. Hudson, and D.E. McKenzie, "Sigmoidal morphology and eruptive solar activity," *Geophys. Res. Lett.*, 26, 1999, pp. 627-630.

2. See articles in N.U. Crooker, J.A. Joselyn, and J. Feynman, eds., *Coronal Mass Ejections, Geophys. Monogr. Ser.*, 99, American Geophysical Union, Washington, D.C., 1997. See especially H.V. Cane, "The current status in our understanding of energetic particles, coronal mass ejections, and flares," pp. 205-215, and D.V. Reames, "Energetic particles and the structure of coronal mass ejections," pp. 217-226. A more recent reference is D.V. Reames, "Particle acceleration at the sun and in the heliosphere," *Space Sci. Rev.*, in press.

3. See the review of shock-acceleration mechanisms by M.A. Lee in Crooker, Joselyn, and Feynman, eds., *Coronal Mass Ejections*, 1997, pp. 227-234.

4. R.A. Leske, R.A. Mewaldt, E.C. Stone, and T.T. von Rosenvinge, "Geomagnetic Cutoff Variations During Solar Energetic Particle Events—Implications for the Space Station," *Proceedings of the 25th International Cosmic Ray Conference*, 2, Space Research Unit, Department of Physics, Potchefstroom University for Christian Higher Education, South Africa, 1997, p. 381.

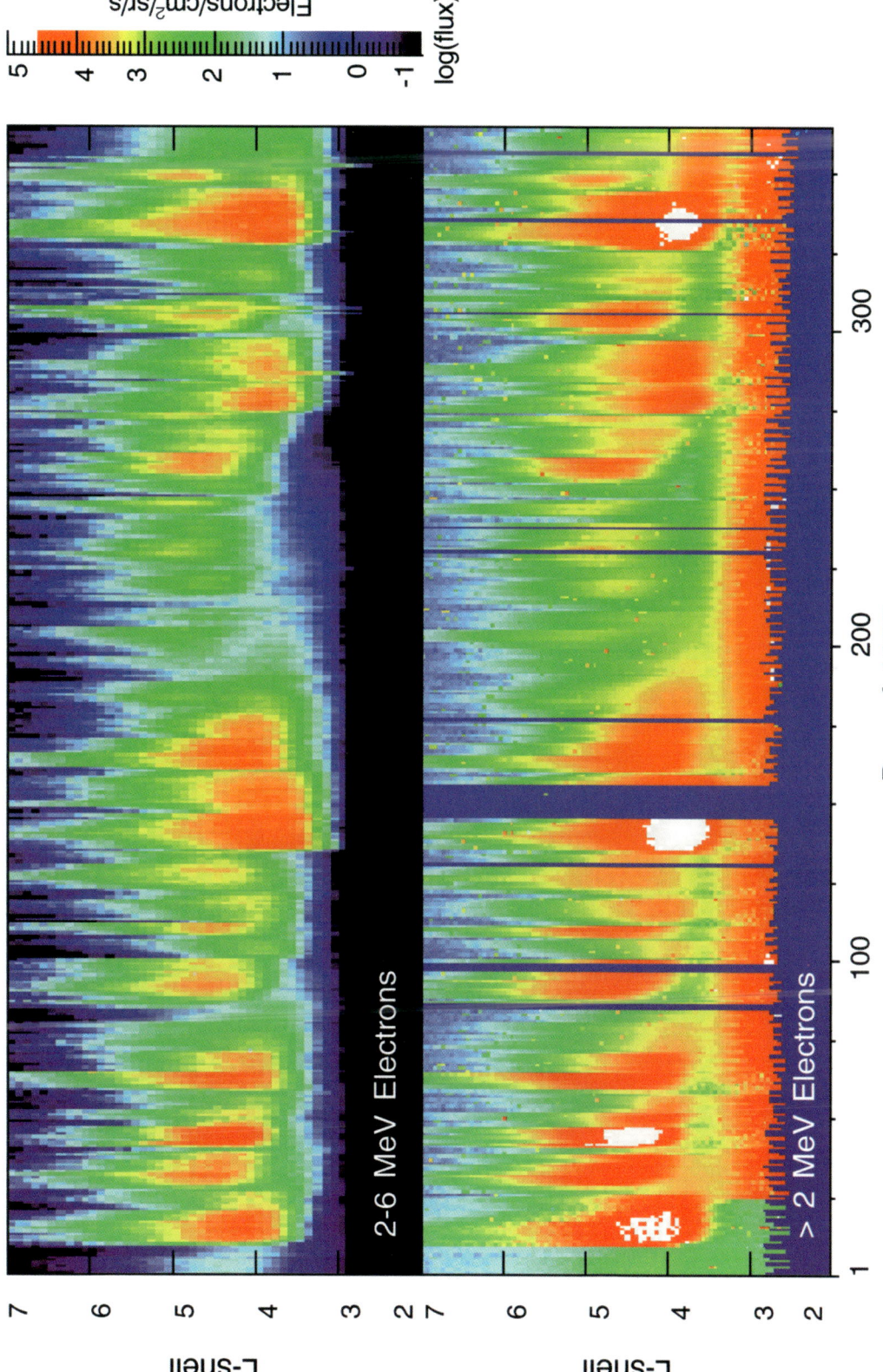

Plate 1 Average flux of energetic electrons during 1997 recorded on SAMPEX (top) and POLAR (bottom) satellites at low (SAMPEX) and high (POLAR) altitudes. Note the appearance on day 20 of a third belt in the POLAR panel near L = 3 following a major magnetic storm.

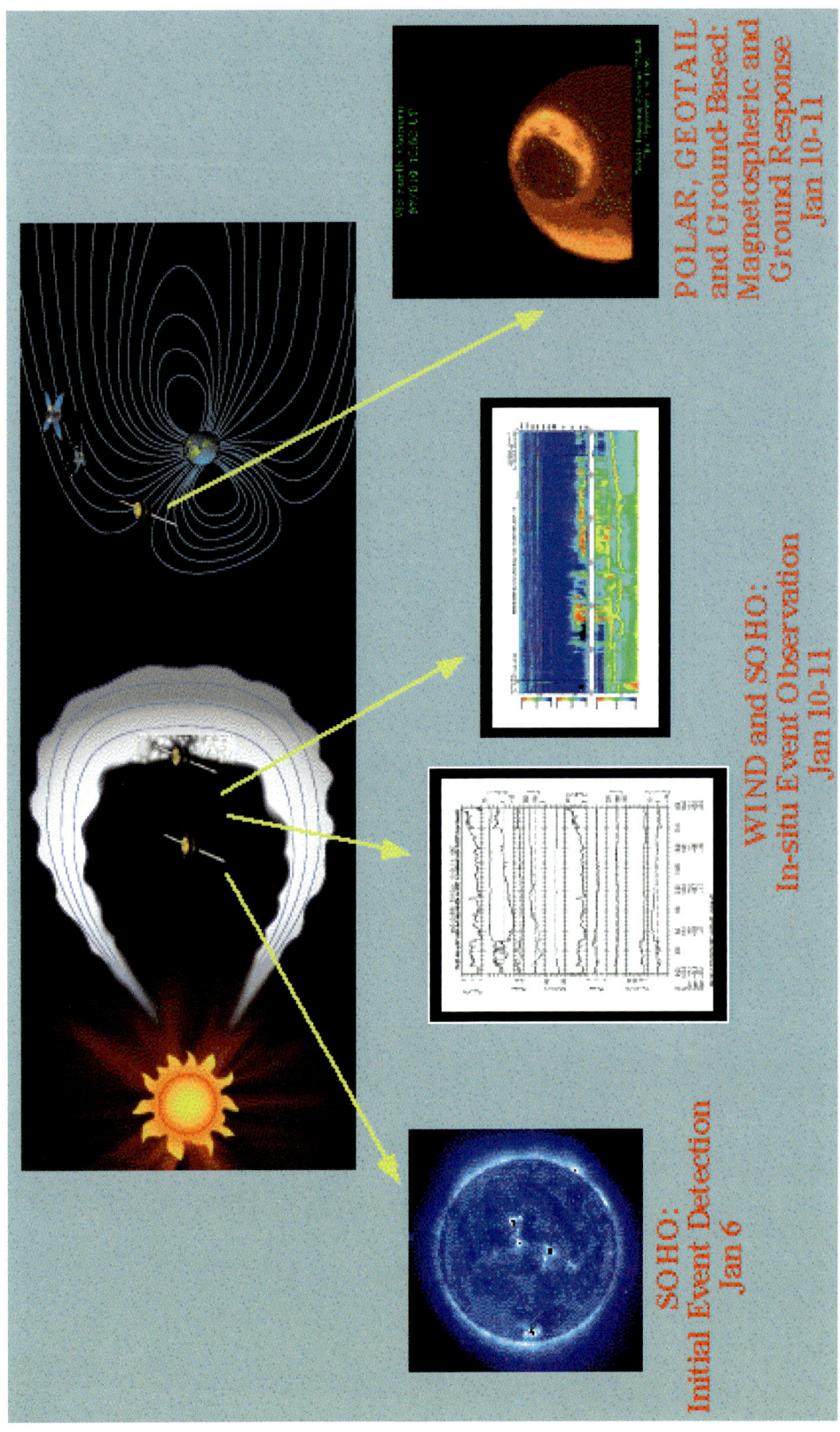

Plate 2 The Sun-Earth Connection event of January 1997. A CME left the Sun on January 6 and passed Earth on January 10 and 11. The figure illustrates the tracking of the event and its effects on the geospace environment.

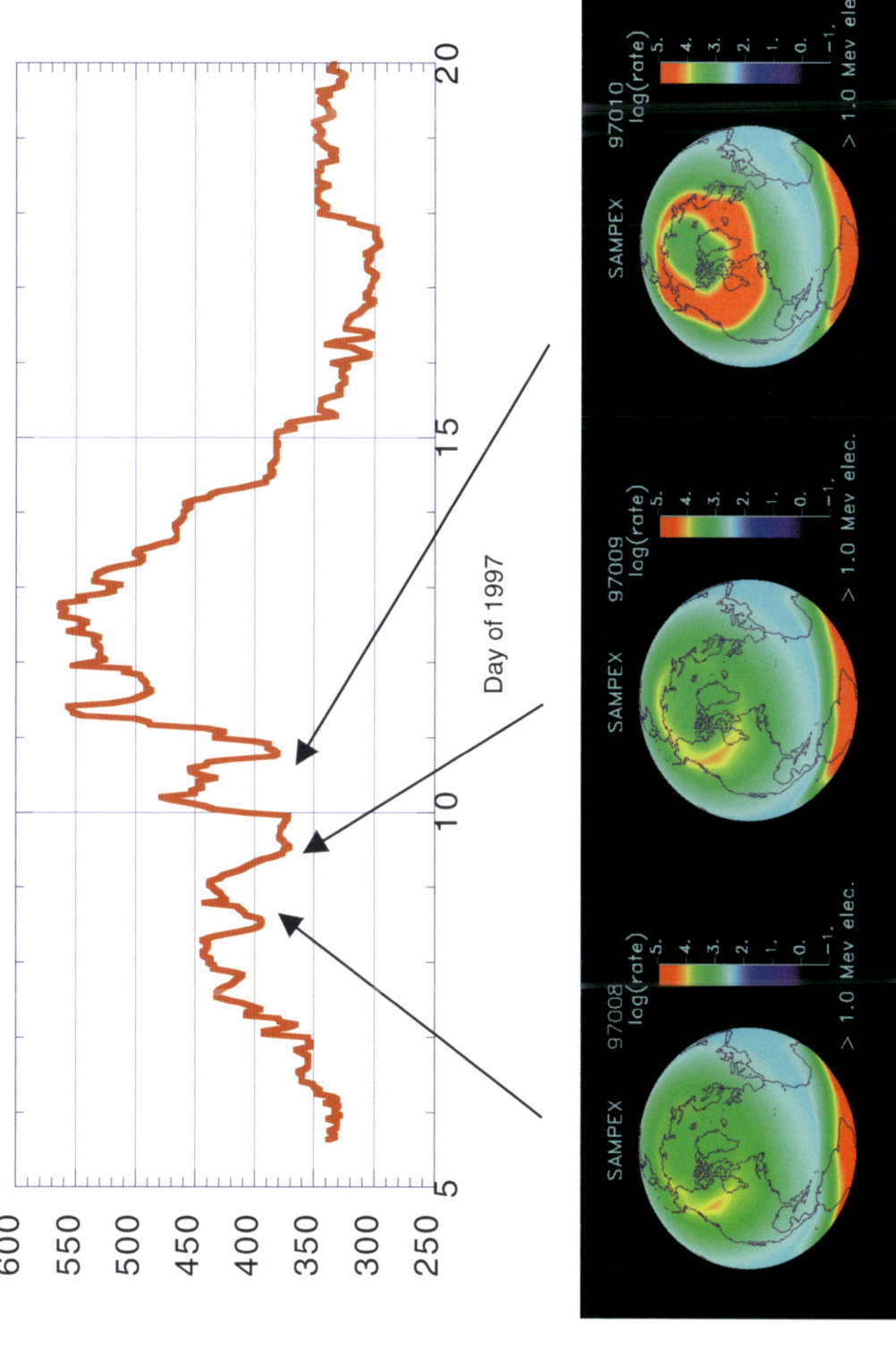

Plate 3 Top panel, solar wind speeds measured on SOHO during the January 1997 Sun-Earth Connection event. Bottom panel, SAMPEX measurements of radiation belt intensities at 600 km altitude. Notice the large increase in SAMPEX intensities on January 10, when a high-speed, high-density stream impacted Earth.

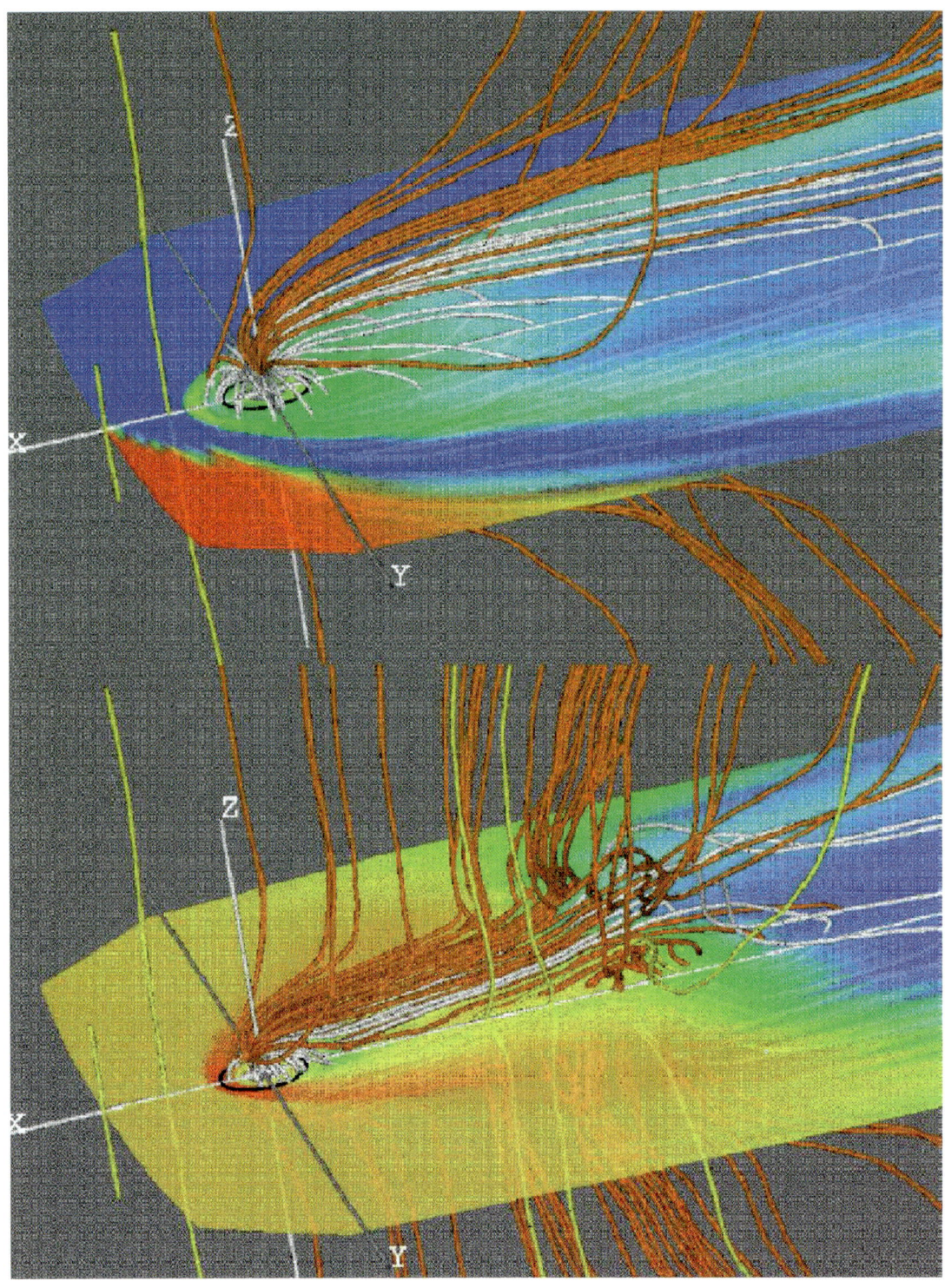

Plate 4 Global MHD simulation of the impact on the magnetosphere of the interplanetary shock wave of March 24, 1991, which violently compressed Earth's magnetosphere and rearranged the radiation belts. The top diagram shows the configuration just before the shock hit; the bottom one shows the configuration a few minutes later, as the shock moved the magnetotail. Area colors indicate temperature in the equatorial plane. Colored lines are magnetic field lines.

5

Interagency Connections

5.1 INSTITUTIONAL FACTORS LIMITING INTERAGENCY ABILITY TO PROVIDE BETTER INFORMATION FOR OPERATIONAL RADIATION RISK ASSESSMENTS

The radiation monitoring and forecasting support provided to the International Space Station (ISS) is limited in its ability to accurately specify and forecast radiation doses at the location of the ISS. Flight operational safety could be improved and operations made more efficient if environmental observations made at other locations could be extrapolated in time and space to the ISS position using a model. Current (unofficial) flight rules require mission operators to base any concrete protective action on actual measurements of the radiation dose on ISS. The risk to ISS crews of exposure to excess radiation could be significantly reduced if other remote environmental measurements could be accurately applied to the ISS environment to provide advance warning of a developing radiation event. (An analogy to the present situation would be a situation in which no steps may be taken to protect people against a hurricane until a wind monitor on the beach registers a high reading.)

The primary obstacle to using remote data is the lack of an accurate modeling capability. As noted throughout this report, models and sensors are available that could significantly improve extrapolation of the radiation dose. However, until they are put in place, information produced from the current space weather system will be unreliable and restricted in usefulness.

Two government organizations, NOAA and NASA, are involved in collecting and analyzing remote environmental data and applying the results of the analysis to ISS operations. The NOAA Space Environment Center (SEC) collects observations of the environment made on satellites operated by NOAA, NASA, DOD, and other organizations. The observations are used to develop a wide-ranging specification of the current environment and a forecast of future conditions. The current specification data and the forecast of future changes are supplied to the NASA Space Radiation Analysis Group (SRAG) at Johnson Space Center. SRAG maintains a running record of the astronauts' exposure to radiation and provides the flight surgeon with information on current and predicted changes in the radiation environment. Flight control teams then use this information to make decisions on flight operations. Both SEC and SRAG could raise the quality of their environmental specification, data analysis, and forecast capability if they were in a position to take advantage of the scientific knowledge that is available from research programs in NASA and in the broader scientific community. Improved specification and prediction of the radiation dose to ISS crew members would lead to improved safety and efficiency.

A major obstacle stands in the way of implementing any such improvements. Both SRAG and SEC have all they can do just to maintain the data collection and analysis that are needed for ongoing operations. The incorporation of improvements becomes a secondary activity, and the lack of adequate resources and agency support in both organizations is limiting the rate of improvement. Currently, the incorporation of improvements lags far behind taking advantage of new capability. Figure 5.1 illustrates the backlog of models that need to be improved before the ISS radiation environment can be accurately specified and forecast.

The transition to better models is virtually stalled by the full commitment of personnel to higher priority operational tasks. Whenever SRAG personnel have the time, they work on integrating improvements. NOAA SEC has introduced a rapid prototyping process in an attempt to overcome the traditionally costly and drawn-out process of "transitioning" models to operational status. This entails subjecting scientific models and results to a linear process of validation and refinement; specifying and designing the user interface; developing mission-qualified software; and "backstepping" the process to make changes necessitated by problems with the science, the validation, or operational or customer requirements. The rapid prototyping process is a dynamic, circular process in which validation, user interfaces, customer products, and software testing are performed simultaneously in the user environment. It involves the scientists who originated the models, the forecasters and other users, and systems support staff working together to reduce the integration time from years to months and to cut costs, which can amount to millions of dollars, by more than 50 percent. The problem with the rapid prototyping process is that even it needs resources beyond the baseline operations staff. SEC has reallocated its limited resources to support the rapid prototyping staff, but the contingencies of day-to-day operations have hampered progress.

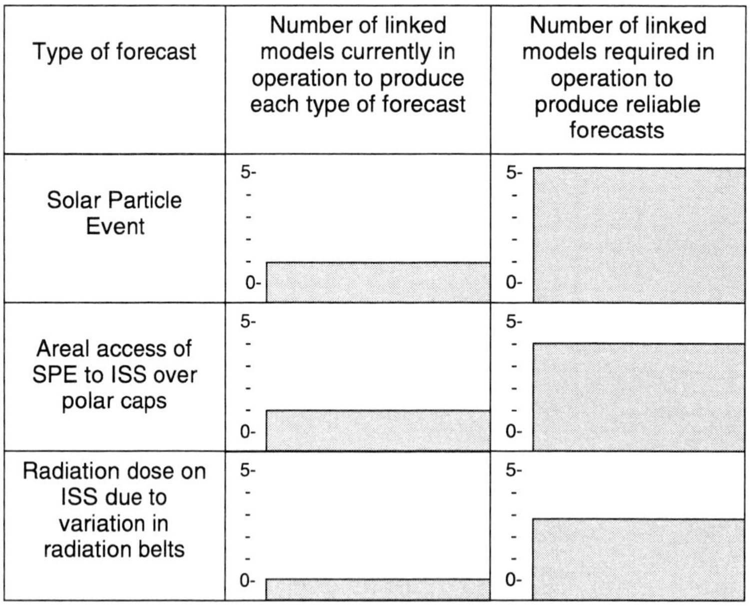

Type of forecast	Number of linked models currently in operation to produce each type of forecast	Number of linked models required in operation to produce reliable forecasts
Solar Particle Event	(low)	(high)
Areal access of SPE to ISS over polar caps	(low)	(high)
Radiation dose on ISS due to variation in radiation belts	(low)	(high)

Figure 5.1 The number of environmental radiation models currently in operation that provide forecasting support to ISS (left column) falls short of the number of models required to provide reliable, accurate forecasts to ISS (right column). The shortfall exists for three types of forecast—forecasting the intensity and time of SPEs (top group), forecasting the area over the polar caps where SPE will reach the ISS (middle group), and forecasting the increase in radiation exposure on board ISS from growth of the trapped radiation belts associated with geomagnetic activity (bottom group).

5.2 RECOMMENDATIONS

Recommendation 5a: NASA, NOAA, and the USAF should cooperate to support the activities that would lead to an operational space weather forecasting capability.

Recommendation 5b: NOAA should extend the range of its SPE predictions from the present ≥10 MeV to biologically effective energy ranges. Forecasts of particle energies at several steps between 10 and 100 MeV would be a significant improvement for space radiation use as well as for other users who operate satellites and systems in space.

6

Intra-NASA Connections

6.1 RADIATION: A CONCERN THROUGHOUT NASA

As the nation's civilian space agency, NASA carries out many activities that require managing the potentially harmful effects of radiation in space. The degree to which these effects must be mitigated is a function of such disparate factors as the potential for human exposure, spacecraft orbit, mission duration, and the sensitivity of spacecraft microelectronics to damage or upset. Consequently, an assortment of space radiation programs and initiatives are distributed throughout the NASA centers and across its enterprises. The priority assigned to these programs also varies; programs related to minimizing the radiation hazard to humans obviously have a very high priority.

While some coordination among these activities exists, CSSP/CSTR believes more would be beneficial. This is reflected in the principal recommendation of this section, which suggests a way to better coordinate NASA's many programs that involve radiation. In the context of this report, the recommendation should be viewed as one element in an integrated program to reduce radiation risk to astronauts involved in ISS construction, maintenance, and operations.

Some units at NASA headquarters and NASA centers have their own programs to cope with the effects of radiation in space, and some units participate in other NASA radiation programs. The programs, several of which are described below, vary according to the area of responsibility of the involved unit. Some relate to engineering applications, others to high-altitude aircraft flights, spacecraft and hardware design, spacecraft operations and reliability, scientific research, or humans in space. At NASA headquarters, programs involving radiation exist in the Office of Life and Microgravity Sciences and Applications (Code U), the Office of Space Flight (Code M), and the Office of Space Science (Code S). Code S, with an emphasis on science, treats space radiation as a phenomenon that it must factor into many of its scientific priorities and objectives, particularly those concerned with major solar disturbances and subsequent geomagnetic storms and disturbances. Code M, with an emphasis on engineering, supports the Space Environment Effects (SEE) program (described below) based at the Marshall Space Flight Center (MSFC). Code U, with an emphasis on biology and medicine, includes NASA's Space Radiation Health Program at headquarters. To convey a sense of the range of diversity of the programs within NASA that involve radiation, a few of the programs are briefly described here.

6.2 NASA PROGRAMS THAT INVOLVE RADIATION

The Space Radiation Health Program (Code U) conducts ground-based radiation studies with appropriate particle species and energy ranges to simulate space radiation effects on cell depletion, on tissue and bone, and on health and living matter in general. These ground-based studies are being carried out at Brookhaven National Laboratory with the expectation that a recent restructuring of the budget will allow the research to be funded and developed in accordance with program objectives. Ground-based research is less expensive than research in space and allows more elaborate and varied experiments to be conducted and repeated. At present, however, not all aspects of the space radiation environment can be simulated in ground-based experiments, especially the synergistic effects of multiple, simultaneous forms of radiation. Thus, ground-based radiation studies must be complemented at least occasionally by well-chosen experiments in space.

The SEE program (Code M), based at MSFC, has close ties to technology and engineering interests at NASA headquarters. SEE is a wide-ranging program: it comprises a number of working groups, and there is participation from other NASA centers, including the Goddard Space Flight Center (GSFC), the Jet Propulsion Laboratory (JPL), and Lewis. Its goal is "to collect, develop, and disseminate the space environment technologies that are required to design, manufacture and operate reliable, cost effective spacecraft for the government and commercial sectors that accommodate or mitigate the effects of the space environment." Towards fulfilling this goal, it provides engineering definitions of the space environment, databases, and design guidelines and attempts to update its capabilities through directed research in response to an occasional NASA Research Announcement (NRA). The SEE program has fostered the development of models of the space radiation environment through its technical working groups, NRAs, and workshops. SEE products, including models and databases, are made freely available to "current and future government and commercial space missions."

One recent SEE study, the Orbiting Technology Testbed Initiative (OTTI) Integrated Trade Study, involved many participants from MSFC and other NASA centers and the DOD and a few participants from the private sector. The objective of the program was to determine the need for and the feasibility of developing a means to test instruments and components for spacecraft that will operate in the high-radiation environment of middle to high altitudes in geospace (above low Earth orbit). Shielding, communications, computer dependability, and the overall reliability and durability of space operations were of central concern to OTTI. The final study recommended that a testbed program be developed utilizing a secondary payload platform to keep costs down. Test flights could begin in a year or two and would be modest in size, mass, and cost, with shared input and objectives from the government and private sectors. Because of its engineering and technology outlook, OTTI did not place scientific needs and requirements among its top objectives, although it did identify science needs. The "trade study" aspect of OTTI refers to the sharing of costs and resources among NASA, DOD, and the private sector.

In October 1998, NASA headquarters adopted a Strategic Program Plan for Space Radiation Health Research (Code U) and designated the Johnson Space Center as the lead center to manage and implement it. The program's goal is to achieve the exploration of space by human beings without subjecting them to unacceptable risks from exposure to ionizing radiation. Implementing the program entails integrating basic science and engineering tools to predict and manage radiation risk for ISS and other deep space exploration projects.

Also located at JSC, the Space Radiation Analysis Group (SRAG) is part of the EVA support team. As was described in Section 1.5, SRAG's responsibilities are to give the flight surgeon the information needed to help protect astronauts from excessive radiation exposure during spaceflight. EVAs are of special concern since an astronaut is shielded less against radiation inside a space suit than inside a spacecraft. SRAG provides preflight projections of the crew's exposure to radiation, real-time estimates of their exposure during flight, and postflight analysis of the radiation actually experienced. For EVAs, a radiation hazard assessment is made in real time, from one hour before EVA egress to its conclusion. SRAG makes go/no go and start/stop recommendations to the flight surgeon, who reports to the flight director.

The radiation program within the Office of Space Science (Code S) concentrates on two themes: the Sun-Earth Connection and Solar System Exploration. Through the years, Code S has accomplished most of the basic research leading to our current knowledge of the space radiation environments of Earth, other planets, and interplanetary space. Of the missions now operating, the most relevant are those in the Mars program and the

International Solar-Terrestrial Physics (ISTP) program, now in its extended phase. ISTP and complementary Sun-Earth Connection missions such as ACE, First Auroral Snapshot Explorer (FASE), and SAMPEX employ an unprecedented fleet of spacecraft to study the Sun, interplanetary space, and Earth's magnetosphere by means of simultaneous observations from different locations in the system (see Chapter 4). For the first time, events can be followed as they develop in space and time, and cause-and-effect relationships can be established. ISTP and the complementary missions were created with solid scientific objectives that continue to be valid. They also show great promise of advancing the scientific understanding of the dynamical space radiation environment of Earth and interplanetary space, which is another reason for extending them. Of particular note, the ISTP theory program has made significant advances in magnetohydrodynamic global modeling of the magnetosphere and its dynamical responses to interplanetary disturbances originating at the Sun (see Appendix A). Since these missions provide the data that relate to the radiation environment, they constitute a common resource for the space radiation activities of the NASA codes and centers. Accordingly, let us turn now to communication across codes and centers.

6.3 COMMUNICATION BETWEEN PROGRAMS WITH AN INTEREST IN RADIATION

A radiation coordinating team involving Codes U, S, R (Office of Aerospace Technology), M, and A (Office of the Administrator) has been established at headquarters to bridge interests across NASA. The team has the authority to conceive and implement projects that cut across codes and centers, but a noticeable absence of such projects suggests that it has been less active than it needs to be to achieve its potential. Instead of coming from the team, cross-code initiatives come from individual project managers. A recent example was the useful discussions and expressions of interest in cooperation between the Office of Life Science (Code U) and the Office of Space Science (Code S), particularly the latter's Sun-Earth Connection Theme and its Solar System Exploration Theme. These discussions resulted in Code U being given the opportunity to conduct Mars radiation environment studies utilizing the Mars Surveyor 2001 and 2003 missions. There are two further areas of endeavor where the coordination of radiation activities within NASA would yield benefits.

Radiation Models for NASA Operations

To satisfy the space weather and engineering requirements of the SEE program, it is necessary to construct better space environmental radiation models than now exist. New models should have a strong scientific basis with regard to both data quality and modeling techniques. A project such as this, which is described in Appendix A, belongs naturally in Code S, but its application is to activities in Code M. A mechanism for cross-code coordination would help here.

Radiation Data for NASA Operations

SRAG confronts severe limitations in trying to fulfill its responsibilities. Although information useful to SRAG may be available from currently operating NASA satellites, none of the mission objectives of these satellites address the needs of SRAG or the issue of human safety in space. It is essentially a matter of luck and not of program priority when a NASA spacecraft becomes available to SRAG. As Chapter 4 details, these spacecraft could be very helpful to SRAG as it tries to fulfill its responsibilities. In the Apollo era, the research and human spaceflight sides of NASA were more tightly coupled. In 1961, the IMP program was planned and "sold" in direct response to the radiation hazard to astronauts in the Apollo program. After the Apollo era, however, and until recently, when the relevance of space weather was emphasized by the Sun-Earth Connection Theme in the Office of Space Science, a relatively low priority was assigned to developing a quantitative understanding of the Earth's space radiation environment. The scientific foundations of this discipline need to be strengthened to facilitate development of the quantitative physical models of disturbed conditions needed by SRAG. Also there is no program or initiative in NASA, such as the Rapid Prototyping Center (RPC) at NOAA (see Chapter 5), that promotes the transformation of physical research models of the space weather environment into operational models such as would be needed by SRAG. The importance of developing operational models for nowcast or forecast purposes has been recognized by the National Space Weather Program (NSWP), but as yet only limited progress has been made, mainly by NOAA's RPC.

6.4 SUMMARY AND RECOMMENDATION

There are major programs at NASA that require an accurate knowledge of Earth's radiation environment. The kind of knowledge required varies from program to program, but the range of knowledge needed extends from the basic science, the physical processes, and the generation mechanisms of the radiation belts and particle events, to net integrated radiation doses averaged over a long period of time. The trend in recent years at NASA has been to smaller and cheaper spacecraft with smaller instruments, more onboard data processing, and increased use of microelectronics. For these and other reasons, the working knowledge of Earth's radiation environment (models, forecasts of particle events and disturbances, integrated doses, etc.) should be improved to address current planning and development requirements in just about every area of NASA activity. A better understanding of the radiation environment, which can come from programs within NASA, would be a great advantage to other NASA projects in science, engineering, technology, and the human spaceflight program. CSSP/CSTR believes the enormous complementary strength in personnel and resources at the different NASA centers should be utilized synergistically to support the needs of both the manned and unmanned programs for information on the space radiation environment.

Recommendation 6: To coordinate intra-NASA activities and concerns in radiation, NASA should establish an agency-level radiation plan and task force. It should also establish a multidisciplinary steering committee to advise the task force.

For greatest effect, the radiation plan should be developed under the leadership of headquarters and with approval of the NASA administrator. The envisioned task force, which could be a revitalized version of the existing radiation coordinating team, would be responsible for managing the radiation plan. CSSP/CSTR suggests that a number of elements be incorporated when implementing this recommendation. The task force would operate across codes. It would report to appropriate high-ranking officials in several offices at NASA headquarters (at the associate administrator level and appropriate program directors or managers). It would prepare an annual report and would be co-led by two or more NASA centers.

CSSP/CSTR suggests that MSFC and JSC be among the centers taking the lead. MSFC is managing the SEE program. JSC is responsible for safeguarding humans in space. CSSP/CSTR suggests that the steering committee be appointed by headquarters and that it have a rotating membership, ensuring representation from other agencies, universities, and industry. It is important that there be representatives on the task force from GSFC and JPL. GSFC representation is important because Goddard has strength in the basic science underlying space radiation, which is crucial for developing and implementing models and designs. For example, the National Space Science Data Center at Goddard could evaluate the scientific merits and inadequacies of current scientific and engineering models and databases. JPL's presence on the task force is needed because JPL is the lead center for deep space missions and so should take part in policy and program developments that involve the interplanetary environment. Also, it has played an important role in developing models of space radiation and space storms.

Epilogue

A Notional Scenario for Improved Support of International Space Station Construction

E.1 VISION OF AN ISS CONSTRUCTION MISSION SUPPORTED BY RELIABLE, ACCURATE RADIATION FORECAST MODELS DURING THE SOLAR MAXIMUM

Present models do not include all of the links that connect sources at the Sun to the radiation environment at the shuttle or the ISS. Moreover, they are primitive and offer only general guidance. Their forecasts of SPE occurrence and size are therefore too inaccurate to cause a flight director to alter a preset schedule of mission activities. Instead, an SPE forecast activates a situational analysis for preparing defensive responses. To illustrate the potential value to NASA of having much better information from space weather services, this section presents a counterfactual situation in which SPE forecasts are accurate enough to cause a flight director to alter a preset schedule of mission activities. One ancillary point made by this illustration is that high solar activity does not necessarily mean high radiation doses. Whereas the ability to make accurate SPE forecasts is usually associated with a decision to postpone or terminate an activity, it could also lead to a decision to initiate an activity. Even when radiation is elevated and further increases threaten, accurate numerical forecast models, which must still be developed and put into operation (see Appendix A), could nonetheless allow a flight director to maintain operations if the models forecast that the environment will remain safe for ISS workers.

E.2 THE WAY THINGS OUGHT TO WORK

It is afternoon on October 15 at the Johnson Space Center in Houston. Inside the control room, a flight director monitors the progress of AS13, the thirteenth U.S. shuttle flight carrying a team of astronauts to construct the International Space Station (ISS). AS13 has 2 full days left, and all has gone well so far. Just one EVA, scheduled for tomorrow afternoon, remains to accomplish mission objectives. But the flight director is worried. A huge active region on the Sun, which produced solar particle events (SPEs) on each of two earlier rotations, is making its third pass across the Sun's maculae-pocked face. Although the region is now approaching central meridian—the most dangerous time for SPEs—no very large events have occurred so far. But space weather forecasters at the NOAA Space Environment Center in Boulder note that a large X-ray sigmoid, a CME precursor, has developed in the center of the maelstrom. They predict that the region will probably erupt in a solar storm in a day or two. Forecasters at NOAA's

SEC and at the NASA Space Radiation Analysis Group (SRAG) console at JSC are communicating. SRAG is following the activity on its monitors in Houston as well as receiving alerts and forecasts from SEC.

What follows is a best-case scenario in which the course of events is controlled by an advanced capability for using space weather forecasts to support ISS construction. It envisages a time when accurate, reliable forecast models—which again, to emphasize the point, must still be developed and "transitioned" into operation—are in place to model the transport of coronal mass ejections from the Sun to Earth, to model the effects of shocks on interplanetary particle acceleration, and to accurately model the interaction of the disturbed solar wind with Earth's magnetic field, especially its effects on the size of the zones accessible to solar energetic particles. Given all this, the history of Flight AS13 unfolds as follows:

15 October, 1730 Houston time (HT), 2330 UT

The Space Environment Center receives data showing a class X flare at S10 W15. The GOES solar X-ray sensor measures a peak flux of 10^{-3} W/m^2, and a time-integrated energy of 0.7 J/m^2 in the 1 to 8 nm X-ray band is accumulated in the first 10 minutes of the event. The solar X-ray imager (SXI) (scheduled launch date: October 2000) registers the event and shows large coronal changes and loop structures, indicating a substantial coronal mass ejection (CME) has occurred. Ground-based solar observatories report that the event includes an optical flare and large radio bursts.

SEC notifies SRAG that the alert criteria for a major X-ray flare and an integrated X-ray event flux have both been reached. SRAG notifies the flight surgeon that a major solar event has occurred.

15 October, 1735 HT, 2335 UT

SEC runs its updated PROTONS 5 Model. It produces a forecast of a large SPE with a 95 percent chance of occurrence. The first increase in 30 MeV protons is forecast to begin in 25 minutes. SEC relays this information to SRAG.

15 October, 1750 HT, 2350 UT

The X-ray event is still in progress. The X-ray-integrated flux reaches 1.5 J/m^2 on the GOES satellite. A new run of PROTONS 5 forecasts a total event fluence of 10^7 p.f.u. with energy (E) > 30 MeV. The peak flux is forecast to occur on 16 October around 0900 UT. SEC reports this information to SRAG. SRAG relays it to the flight surgeon and the flight director.

15 October, 1825 HT, 16/0025 UT

Energetic particle sensors on GOES begin to show an increase in energetic particle fluxes over the range 10 to 100 MeV.

15 October, 1840 HT, 16/0040 UT

The on-board radiation instrumentation, including the tissue equivalent proportional counter (TEPC) and the internal vehicle charged-particle directional spectrometer (IV-CPDS), which have been temporarily moved about the habitable cabin for a background radiation survey, are to be returned to their usual location. The TEPC alarm is turned on and set at a low threshold, and real-time telemetry from all radiation subsystems is activated.

15 October, 1855 HT, 16/0055 UT

The Large Angle and Spectrometric Coronagraph Experiment (LASCO) coronagraph data, combined with radio and plasma wave data from the WAVES instrument on the Wind spacecraft, are used as input to the advanced Shock Propagation Model (SPM). Its output includes the prediction of a shock arriving at Earth between 1200 and 2000 HT on October 16. It also includes a forecast of B field structure and shock strength. This information is used to run the advanced SPE Profile Forecast Model (SPEPFM). The model forecasts that SPE fluxes will remain below 500 p.f.u. for particles with energies of 10 to 30 MeV until about 16/1600 HT. At that time, the particle fluxes are predicted to begin to increase and to rise by two orders of magnitude over the following 4 to 5 hours. The version 3.0 Neural Net Fluence Model (NNFM) predicts an integrated fluence of 10^8 to greater than 10^9 p.f.u. with E > 30 MeV for the entire event. SEC reports this information to SRAG.

15 October, 1930 HT, 16/0130 UT

Using input from SOHO, SEC refines the Shock Propagation Model forecast to an arrival time of 16/1630 HT (16/2230 UT). The advanced Polar Cap Model predicts an opening of the polar caps down to 48 degrees geomagnetic latitude. The NNFM forecasts the time-integrated particle flux to exceed 10^9, with less than 10 percent error in the forecast.

15 October, 2000 HT, 16/0200 UT

SRAG reports the SEC forecasts to the flight surgeon and the flight director. A morning rest period is scheduled for the astronauts on the following morning and a 6-hour EVA is to begin on October 16 at 1300 HT. All three parties confer. SRAG reports the doses on an EVA for these fluxes will be low through the morning of the 16th but will be much higher in the afternoon, after 1600 HT. The flight surgeon begins to review the radiation records of individual crew members.

15 October, 2300 HT, 16/0500 UT

Telemetry is live from the TEPC, EV-CPDS, and IV-CPDS. SRAG compares the observed dose profiles with the projected doses from satellite particle data and finds them to be within 15 percent agreement. SRAG and SEC review the observed SPE profile, which is within 20 percent of the forecast from the SPEPFM.

16 October, 0200 HT, 0800 UT

After conferring with the flight surgeon and SRAG, the flight director decides to move the EVA to 0900 the following morning.

16 October, 0800 HT, 1400 UT

Spacecraft dosimeter readings and dose calculated from energetic particle data are in agreement. The dose rate is low. Telemetry continues live from TEPC, EV-CPDS, and IV-CPDS. In case of space telemetry shutdown because of the effects of the SPE on system hardware, doses calculated from satellite SPE observations will be used on the ground to monitor the crew dose. The EVA begins.

16 October, 1053 HT, 1653 UT

The TEPC alarm sounds. After consultation between the flight director, the flight surgeon, and SRAG, it is determined that although the dose rate has gradually reached the alarm level it is still low in terms of crew exposure. SRAG consults with SEC and decides the rapid increase is still not expected before 1600 HT and $5^1/_2$ hours are left before the dose rates rise rapidly.

16 October, 1200 HT, 1800 UT

Satellite particle fluxes begin to increase rapidly at 1430 UT. They reach 15 times the background level before the increase. Doses inside the shuttle increase even more rapidly because of the hardening of the event spectra. The crew goes to the docking module for additional protection during the five polar cap passes that impart the most exposure.

16 October 1330 HT, 1930 UT

The dose rate has continued to increase slowly but remains low as measured by the on-board dosimeters being monitored on the ground by mission control. The EVA is completed 30 minutes early and the crew returns immediately to the spacecraft.

16 October, 2400 HT, 17/0600 UT

SPE fluxes and dose rates return to the levels of early on October 16. Reentry is still as scheduled, the next morning.

By using a (hypothetical) full-up space weather capability, with integration of advanced forecast modeling capabilities, the ISS has completed its mission and avoided an extra day in orbit, which would have been necessary if the last EVA had been carried out as originally scheduled. In that alternative case, the EVA would probably have begun and then been cancelled as the radiation levels rose or the crew would have suffered excessive radiation doses while trying to complete the EVA.

E.3 THE MISSING PIECES

A number of advanced models are posited in this scenario:

- Updated PROTONS 5 Model, which predicts SPE from solar X-ray inputs;
- Advanced Shock Propagation Model, which predicts shock arrival time and post-shock parameters;
- Advanced Profile Forecast Model, which predicts flux profiles as the event proceeds for a standard set of energies;
- Version 3.0 Neural Net Fluence Model, which predicts integrated fluence for the event; and
- Advanced Polar Cap Model, which specifies energetic particle access zones from predicted postshock and CME conditions.

All of these models, which have been given descriptive hypothetical names, are in various stages of development. None is up to the task assigned it in this scenario, but as Appendix A describes, each is a goal of the National Space Weather Program (NSWP) and of scientists involved in space weather modeling. The models will not be ready in time to make the scenario described here a reality during the planned construction phase of ISS. NSWP and the community of space weather scientists and service providers intend, however, to improve the state of space weather services by the time the operational phase of ISS has been reached so that it approaches the level of capabilities assumed for these ideal models.

E.4 TIMETABLE FOR IMPLEMENTING THE REPORT'S RECOMMENDATIONS

A variety of actions are recommended in this report to support the construction of the ISS during a solar maximum. Figure E.1 gives a timeline keyed to the ISS construction schedule (as it was known in July 1999) for the implementation of the recommendations of this report. As shown in the figure, there are recommendations whose implementation should be completed right away (R1, R3a, R4, R5b, and R6); recommendations that will require one or two years to implement (R2 and R3b); and recommendations that will take several years to implement (R3c and R5a). The research needed to improve space weather services in support of manned missions (described in Appendix A) is also shown on this figure. Research activities are organized into two groups: those that can be implemented within 5 years (A1), with some activities being implemented within 1 year, and those requiring more time to implement (A2).

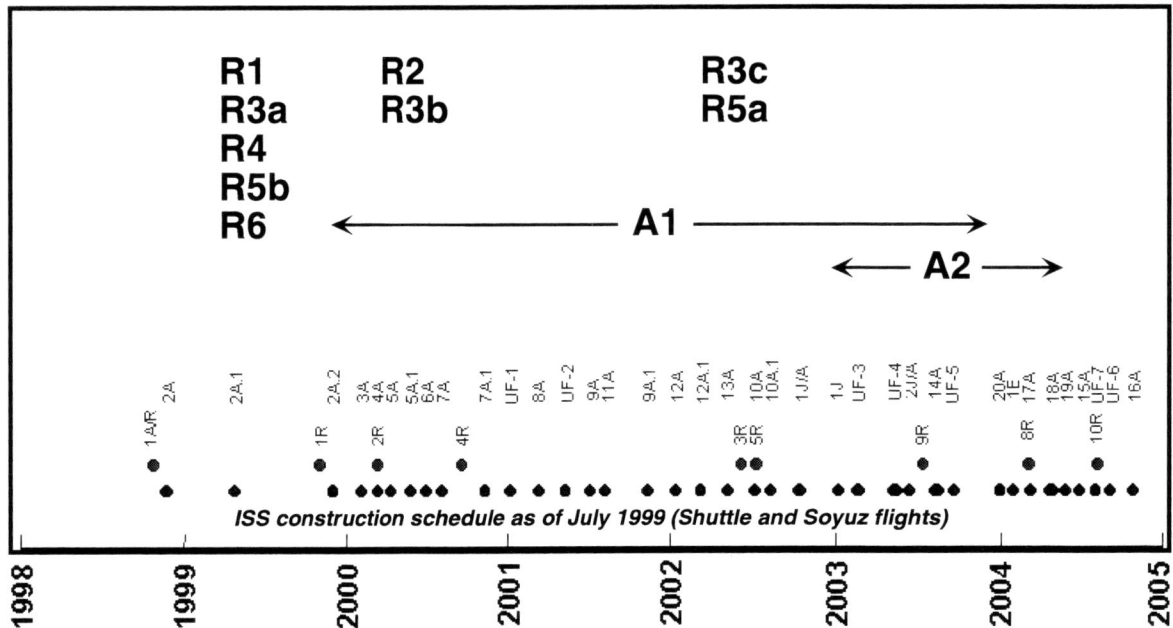

Figure E.1 Timeline for implementing the report's nine recommendations, which are denoted R1 through R6. A1 and A2 refer to priority research activities (see Appendix A). The timeline is keyed to the ISS assembly schedule, which is available from NASA on the Web at <http://spaceflight.nasa.gov/station/assembly/flights/chron.html>. The first crew—a U.S. astronaut and two Russian cosmonauts—will be launched on a Russian Soyuz spacecraft in March 2000 on flight 2R ("R" denotes a Russian mission). They will stay there 3 months. From that point on, the ISS is planned to be permanently staffed.

Appendixes

APPENDIX A

Space Weather Models Applied to Radiation Risk Reduction

A.1 SPACE WEATHER MODELS

To optimize our ability to avoid radiation risks to humans in space, a system should be developed that can provide "timely, accurate, and reliable space environment observations, specifications, and forecasts."[1] At present, there are significant gaps in our understanding that diminish the ability of space weather models to perform this function. Nonetheless, the science of space physics has matured to the point where it is able to describe and model many aspects of the complex links between the Sun, the interplanetary medium, and Earth. These capabilities are essential for predicting space weather in general and the radiation environment at low Earth orbit in particular. Recent advances in numerical technology and computer architectures have meant a rapid growth in our ability to model space weather, in particular the propagation of solar eruptions through the heliosphere. The last few years have also seen rapid advances in our capabilities for representing the present state of the magnetosphere and providing short-term forecasts, and progress is likely to accelerate in the next few years.

The presently available empirical and semiempirical models that have the potential for providing predictions that will be useful during the ISS construction period should be adapted for operational use. CSSP/CSTR notes at the outset one criterion bearing on the likely usefulness of a model. From a flight director's perspective, false alarms are worse than missed events. Models with low false alarm rates are therefore the ones most likely to give predictions that flight directors would trust enough to act upon.

A.1.1 Solar and Inner-Heliospheric Models

CMEs and flares are the first links in the chain of efficient causes that connects eruptions on the Sun to space storms at Earth, including SPEs. To understand the solar origins of space weather, therefore, we must understand how CMEs and flares are initiated. The following discussion will focus primarily on fast CMEs, the predominant source of SPEs and geomagnetic storms. At present, we know the following:

- The Sun's magnetic field is the most likely source of the substantial energy needed to launch and maintain a CME.
- Most CMEs come from the streamer belt or from the boundary between the polar coronal holes and adjacent active regions.

- CMEs are not caused or driven by flares; in some cases the causal relationship appears to be reversed, but in many cases the relationship appears more complex.
- Fast CMEs and some slow CMEs drive shocks, which in turn accelerate particles.

CME initiation is still poorly understood. Existing theories of CME initiation generally can be categorized according to the underlying driver: gas-pressure-driven models, including buoyancy and flare-driven mechanisms; ideal magnetohydrodynamic (MHD) models, based on the emergence of subphotospheric flux ropes into the corona and their subsequent loss of equilibrium (for example, through prominence draining, which removes "ballast" holding down the flux rope); and resistive MHD models, which invoke either the formation and detachment of a flux rope through reconnection underneath (the so-called tether-cutting process) or the removal through coronal reconnection of magnetic flux above a stressed configuration, which then is free to erupt (the so-called breakout model).[2-5] Most CME initiation models have only been studied analytically or through two-dimensional simulations. The use of solar observations to establish boundary conditions in such calculations and the move to three-dimensional simulations have both become feasible only with the advent of high-performance computers.

Modeling the propagation of fast CMEs and their associated shocks through the heliosphere, even with one- and two-dimensional MHD codes, has reached the point of demonstrating measurable skill in predicting shock speeds and arrival times at Earth.[6] Recent improvements to these models include the incorporation of three dimensions, the ambient solar wind structure, and the heliospheric current and plasma sheets.[7] Once an eruption has begun, it is a relatively straightforward matter to simulate its propagation as that of a pulse travelling through the heliosphere, as long as the plasma and magnetic-field conditions along the way are known. Unfortunately, this requirement becomes increasingly difficult to satisfy at and after a solar maximum, when close sequences of CMEs or eruptive flares and a highly distorted heliospheric current sheet prevent the establishment of a predictable, quiescent interplanetary medium. Real-time monitoring of the ambient interplanetary medium properties to establish background conditions would improve the accuracy of CME and shock propagation models. In addition, we do not know where these shocks first develop, which is a key determinant of the particle acceleration process addressed below.[8] Better observations of, and insight into, the sources of type II solar radio emissions, which give information on shock speeds, would be valuable for initializing and testing models. Finally, the relationship between CMEs, which are identified as solar phenomena, and magnetic clouds, which are observed as interplanetary structures, is an unresolved puzzle that merits increased attention by both observers and theorists. Despite these caveats, the heliospheric phase of space weather development is reasonably well understood. Consequently, models of interplanetary propagation hold great promise for supporting NASA's efforts to keep astronauts from being exposed to harmful levels of radiation during ISS construction and beyond.

The next link in the Sun-Earth chain is the acceleration of particles by CME shocks. Several theories exist for this process,[9] quantitative modeling of which provides the only feasible way to connect the MHD characteristics of the CME-shock system with the flux and spectrum of SPEs at Earth. This is a challenging task, as both large-scale MHD characteristics and small-scale particle properties must be considered. A related problem is our incomplete understanding of the flare-driven component of SPEs. Observations by SOHO, Wind, and other spacecraft are clarifying some of these complex phenomena.

Now and in the foreseeable future, only MHD models can span the enormous distances and range of scales from the Sun through the heliosphere into the magnetosphere. A principal long-term goal of the National Space Weather Program (see Section A.5) is to develop a three-dimensional, multiscale model of the heliosphere-magnetosphere-ionosphere system extending from the base of the solar corona to the base of the ionosphere. In its fully developed form, this ultimate space weather model will solve a set of multispecies, generalized, time-dependent MHD equations and will self-consistently describe the complicated interplay among the physical processes controlling the structure and dynamics of the heliosphere and the geospace environment, including the solar wind outflow, the generation and propagation of transient interplanetary structures, such as CMEs or corotating interaction regions (CIRs), and their interaction with the magnetosphere-ionosphere system.

A.2 NEAR-EARTH SPACE ENVIRONMENT MODELS

In the Sun-to-Earth flow of space weather influences, the Sun is the producer, the solar wind the deliverer, and the magnetosphere the receiver. The preceding section dealt with production and delivery. This section deals with reception.

A.2.1 Magnetospheric Conditions Influencing SPE Penetration

As shown in Chapter 2, for several orbits on most days ISS enters the high-latitude regions, which during SPEs are filled with solar energetic particles. Accurate forecasts of astronaut safety therefore depend critically on accurate estimation of the physical extent of solar particle penetration. Among the forecaster's tools are maps of the equatorward edge of the auroral oval. In addition, forecasters can calculate the "solar particle cutoff." The ability of energetic particles to reach a specified point (latitude, longitude, and altitude) above the Earth can be characterized by the solar particle cutoff, which is controlled by a particle's energy (strictly speaking, by its rigidity, i.e., its momentum per charge). Solar particles with energies above the cutoff value can penetrate to the given location, whereas less energetic particles cannot. Although cutoffs traditionally have been calculated without considering the instantaneous state of the magnetosphere, we now know that the SPE zones grow when the magnetosphere is compressed by a CIR or a solar storm or inflated by a strong storm-time ring current.

A.2.2 Dynamic Physical Models of the Radiation Belts

Since their development in the 1960s and 1970s, NASA radiation belt models AP8 and AE8[10, 11] have been very widely used to provide quantitative representations of the average observed particle fluxes. Only long-term space weather effects have been included, primarily by the creation of separate Solar Max and Solar Min NASA models. A new set of empirical models, including the electron model CRRESELE and the proton model CRRESPRO, was developed recently by the Air Force Research Laboratory based on measurements from the CRRES mission of 1990-1991.[12, 13] CRRESPRO is separated into quiet, active, and average models. The quiet model is taken as an average over the ~7 months preceding the great storm of March 1991, and the active model is an average over the following ~7 months, including the secondary proton belt formed by the March storm. The average model is the average over the entire 14-month mission. The CRRESELE model consists of six models ordered by the 15-day average of the geomagnetic activity index Ap (a relative of the Kp index mentioned in Section 2.3), plus mission-average and worst-case models. For certain radiation environment studies, direct dose measurements are preferred to flux measurements because of the uncertainty involved in transforming one to the other. CRRESRAD is a directly measured dose model constructed from CRRES data that are sorted by the same time intervals as CRRESPRO;[14] APEXRAD is a low-altitude dose model constructed from APEX satellite data and sorted by geomagnetic activity in a manner similar to that used for CRRESELE.[15] Although these empirical models take some account of various levels of geomagnetic effects, they obviously do not predict the detailed response of the radiation belts to variations in the magnetospheric configuration, as would a physics-driven model.

On the theoretical side, radiation-belt models traditionally assume adiabatic particle motion, with deviations characterized by diffusion coefficients chosen to fit the observed fluxes. This approach has been very effective in interpreting the average or quiet-time configurations of the radiation belts, but it cannot represent the dramatic flux increases due to solar storms, which pose the greatest risk to astronaut safety. Among the features that are not well described by the diffusion codes are the rapid changes that occur frequently in the outer-belt MeV electrons. For example, these particles typically disappear rather suddenly at the beginning of a geomagnetic storm caused by a CIR, remain at a low level from the main phase through the early recovery phase, and then return with flux levels sometimes orders of magnitude higher than the prestorm values. As discussed in Chapter 3, the decay rate of these enhanced electron fluxes can be estimated once the peak has been observed, but we cannot yet reliably estimate the timing or intensity of the peak. The physics governing the peak flux of these outer-belt electrons is not yet understood but is the subject of intense research activity (see Chapter 3 and D.N. Baker et al.[16]).

A dramatic case of nondiffusive behavior occurred on March 24, 1991: a remarkably strong interplanetary shock smashed into Earth's magnetosphere, briefly bringing the magnetopause to about 3 Re above the terrestrial surface and creating a new radiation belt between the inner and outer belts. In fact, the changes in the radiation environment wrought by this event were so substantial and long lasting that, as noted above, two sets of CRRES-based empirical radiation-belt models were developed to represent the averages before and after this event. The main features of these radiation-belt disruptions were quickly and convincingly explained in terms of a theoretical model that followed gradient-drifting particles in a time-varying magnetic field, first using an ad hoc quantitative representation and later a realistic, three-dimensional MHD simulation of the sudden compression (Plate 4).[17] Note, however, that this simulation followed equatorially mirroring particles through only a few minutes of magnetosphere time.

A.2.3 Real-Time, Data-Driven Map of the Radiation Belts

The number of spaceborne instruments currently monitoring the radiation belts in real time is sufficient to provide the desired comprehensive picture of the belts if the data from these spacecraft can be effectively merged. Prime candidates include the Comprehensive Energetic Particle and Pitch Angle Distribution (CEPPAD) instrument on NASA's POLAR spacecraft, particle detectors on monitoring spacecraft operated by NOAA and Los Alamos National Laboratory, and the simple particle detectors on several Global Positioning System spacecraft. If these data can be optimally merged, making use of the known coherence characteristics of the radiation belts, a useful real-time map of the condition of the belts could be constructed.[18]

A.3 ADVANCED EMPIRICALLY BASED FORECAST MODELS OF RADIATION RISK PARAMETERS

Neural net and nonlinear dynamics models of radiation risk parameters may provide the quickest way to bring forecasting capability to radiation risk management.[19] A large community of researchers is working in the area of applied empirical modeling. Neural nets have been trained to perform diverse forecasting tasks, including the following:

- Make short-term flare predictions with a success rate reported to be around 80 percent;[20]
- Predict the total dose of an SPE, reportedly to within 20 percent, from the fluxes observed earlier in the event;[21]
- Predict up to an hour ahead geomagnetic disturbance indices from L1 solar wind and interplanetary magnetic field (IMF) data,[22] which can then be turned into predictions of the size of the SPE zones (see Section 2.3); and
- Predict the intensity of relativistic electrons at geostationary orbit a day ahead, based on a series of past Kp values, reportedly with an efficiency better than 90 percent.[23]

The evidence suggests that models based on neural nets and nonlinear dynamics can be developed that measurably reduce the radiation risk entailed in ISS construction and operations. A way is needed to focus and coordinate the development efforts on such a project.

A.4 OBSERVATIONAL SUPPLEMENTS TO MODEL-BASED FORECASTS

One approach to improving such forecasts lies in monitoring for "halo" events—CMEs directed at or away from Earth, which appear as expanding annuli or disks of enhanced density roughly centered on the Sun. Although these events were discovered by the SOLWIND coronagraph (SOLWIND) two solar cycles ago,[24] the coronagraphs currently operating on SOHO's LASCO experiment are the first to observe a significant number of halo CMEs, thanks to their extended field of view and their improved sensitivity compared with earlier coronagraphs.

Because coronagraphs are constrained to observe limb events best, however, their data alone do not distinguish those CMEs aimed toward Earth from those originated on the far side of the Sun. The source region can be derived best from concurrent extreme ultraviolet observations of the lower corona using, for example, the Extreme Ultraviolet Imaging Telescope (EIT)/SOHO, which is sensitive to the high-temperature plasmas and transient dimmings characteristic of eruptive events. Vector magnetograph observations (and extrapolations into the corona) taken a few hours before CME onset would roughly indicate the orientation of the CME magnetic field at its origin, providing one factor that contributes to the extent of southward IMF at Earth. Closer to Earth, IMF monitors (e.g., ACE, SOHO, Wind, and IMP 8) can provide the missing link in magnetic topology; this would diminish considerably the lead time between a CME-based warning and the onset of geomagnetic effects and would also reduce the number of false alarms.

A striking illustration of the value of this approach is given by recent studies comparing LASCO halo events with large geomagnetic storms. In 1995, the accuracy of such forecasts was poor: only 27 percent of 173 observed storms were forecast correctly, while 63 percent of the 126 forecasts were false alarms.[25] In contrast, the study of 25 front-side halo CMEs seen by LASCO and EIT during 1996 and 1997 revealed that over 85 percent portended geomagnetic storms with Kp greater than or equal to 6, and only 15 percent of such storms were not predicted (C. St. Cyr, personal communication, 1999). Note that no solar magnetograph or in situ IMF data were used in this study, so it is likely that fewer false alarms would have been reported if this context information had been available. Further work is needed to test and refine this procedure with a larger data set and to determine the cause of the geomagnetic storms, which are especially worrisome because they are apparently unaccompanied by detectable halo CMEs. In the meantime, this promising approach can and should be implemented by NASA to lessen radiation hazards during ISS construction.

Radio-frequency emissions associated with type II bursts give information on the outward propagation speed of the ejected plasma, in particular, providing important clues as to the origin and speed of shocks both close to the Sun and in interplanetary space. However, type II bursts that are first detected in the low corona are primarily associated with impulsive flares, whereas interplanetary type II emissions are clearly associated with CMEs. In addition, the mechanism whereby the shocked plasma produces type II emissions is poorly understood. Finally, some type II emissions have been detected during slow CMEs, which in theory should not be able to drive shocks. Until these uncertainties surrounding the nature of type IIs are clarified, their usefulness as a proxy or early warning of oncoming solar storms remains compromised.[26]

The majority of SPEs are associated with CME-driven shocks, which take a day or two to reach us, thus in principle allowing plans for astronaut activities to be altered as needed. The problem is that unless measurements are made of the source region and the magnetic field geometry of the approaching shock/CME system, as well as of the white-light signature, many false positives will be predicted, canceling out the advantage of the long lead time. Similarly, if forecasters do not detect most of the CMEs aimed at Earth, the relativistic shock-accelerated particles unfortunately will be the sole "warning" of the oncoming geomagnetic storm. Continuous monitoring of the Sun using SOHO-type instruments could provide early warning for many if not the majority of SPEs and geomagnetic storms.

A.5 NATIONAL SPACE WEATHER PROGRAM

The National Space Weather Program, which is managed by the National Science Foundation (NSF) but which also integrates and coordinates the space weather interests of five other agencies—NOAA, USAF, NASA, DOI, and DOE—provides an institutional structure that could help implement the actions suggested below. In broad terms, the goal of the NSWP is to improve the ability of the nation's providers of space weather services to nowcast and forecast the space environment accurately. As the following quote from NSWP's Strategic Plan makes clear, many development areas discussed in this section fall within its purview: "The National Space Weather Program encompasses all activities necessary for the timely specification and forecast of natural conditions in the space environment that may impact technical systems. The domain of primary interest to the program includes the sun and solar wind, the magnetosphere, the ionosphere, and the thermosphere. Because of the vastness

Table A.1 Space Weather Parameters and Goals

Space Weather Domain	Goal
Solar coronal mass ejections	Specify and forecast occurrence, magnitude, and duration
Solar activity flares	Specify and forecast occurrence, magnitude, and duration
Solar and galactic energetic particles	Specify and forecast at satellite orbit
Solar wind	Specify and forecast solar wind density, velocity, magnetic field strength, and direction
Magnetospheric particles and fields	Specify and forecast global magnetic field, magnetospheric electrons and ions, and strength and location of field aligned current systems. Specify and forecast high-latitude electric fields and electrojet current systems
Geomagnetic disturbance	Specify and forecast geomagnetic indices and storm onset, intensity, and duration
Radiation belts	Specify and forecast trapped ions and electrons from 1 to 12 Re

NOTE: This table repeats a portion of Table II.2 in the NSWP Strategic Plan (see note 1).

and complexity of the region of interest, all traditional areas of space sciences can contribute to achieving the program goals."[27]

The approach the NSWP is taking to achieve its goal is also compatible with the development of models that can be applied to reducing radiation risk to ISS construction and operating teams. It calls for developing both data-based models for the near term and deterministic models for the middle and long term, consistent with this passage from the Strategic Plan: "In order to substantially improve forecasting abilities, several advancements must take place. We can make some, albeit limited, progress without improving our understanding of the physical mechanisms which generate space weather. This will require examining and applying data to develop or improve statistical or empirical models. However, in parallel, as our understanding of the space environment improves, physics-based research models will be developed and modified as part of the process of improving our understanding."[28]

The Strategic Plan specifically calls for developing nowcast and forecast models in areas of direct importance for managing radiation risk for ISS construction and operating teams. Table A.1, which is an excerpt from Table II.2 in the NSWP Strategic Plan,[29] shows that the models needed for ISS radiation risk management form a subset of models whose development is called for by the NSWP. The priorities advocated in this Appendix are, therefore, addressed to the affiliated agencies of the National Space Weather Program.

A.6 SUMMARY AND FINDINGS

Space weather modeling aims to develop models that use information from places where instruments happen to be to specify and forecast conditions at places where the information is wanted. Instruments are deployed in space and on the ground in arrangements intended to optimize their usefulness. Nonetheless, there are strategic holes in instrument coverage that planned missions will help fill, as discussed in Chapter 4. On the modeling side, the situation can be described as encouraging. Physics-based, MHD-type models have progressed to advanced stages in all the links that connect solar storms to terrestrial effects. Data-based, neural net, nonlinear filtering models appear to be close to producing operational-quality forecasts of radiation risk parameters.

The CSSP/CSTR findings address the specific elements of an effort that would lead most directly to reducing radiation risk by providing high-quality information on the parameters most crucial to assessing radiation risk. These findings identify and prioritize two kinds of project. The first kind includes relatively mature projects of central importance to radiation risk management that will provide results soon. It also includes projects that will ultimately provide comprehensive (end-to-end) information relevant to radiation risk management and that therefore merit early institutional emphasis. The second kind includes projects that are less mature: although they will ultimately be of great value, it is likely to take longer to realize the results of their operational use in radiation risk management.

A.6.1 Projects Deserving Earliest Possible Attention

Because the following projects could, if started now, be completed in time to help reduce radiation risks during ISS construction and operations, they deserve the earliest possible emphasis by one or more of the affiliated agencies of the National Space Weather Program:

1. Critically evaluate and develop the best of (at least) the following methods that have been suggested for mapping latitudinal cutoffs for SPE particles at the altitude of ISS:

   • *Numerical integration of particle trajectories, using semiempirical models of the magnetosphere (e.g., Tsyganenko[30]).* Results should be compared with observations to assess the degree of improvement offered by this approach.
   • *Direct inference from real-time monitoring of magnetospheric boundary locations.* For example, magnetometers on board the Iridium spacecraft should be capable of mapping the equatorward edge of the auroral oval. Solar protons are observed to penetrate to lower latitudes when auroras reach lower latitudes, but more research is needed to define the quantitative relationship between the location of the edge of the auroral oval and solar-particle cutoffs.
   • *Numerical integration of particle trajectories, using three-dimensional global MHD models of the magnetosphere driven by real-time solar-wind data upstream from Earth.* In the near future, it will be technically feasible to run a magnetospheric model of this type in real time at a forecast center, thus improving short-term (up to about an hour ahead) forecasting. Such global MHD models provide the only means for obtaining a realistic representation of the magnetosphere under extreme conditions, e.g., in response to the impact of an interplanetary shock followed by a CME.

2. Several existing advanced, data-based space weather nowcast and forecast codes could be developed relatively quickly into operational codes that can give SEC and SRAG the ability to forecast at least some radiation-risk parameters during most of the ISS construction period.

3. A prototype of a comprehensive space weather simulation tool could be developed, tested, and made available to a forecast center in time for the peak radiation hazard during ISS construction, given sufficient investment of resources in the near future. Cooperation among the concerned agencies—NASA, NOAA, DOD, and DOE—would be essential for cost-effective and timely progress. By using measurements of the solar magnetic field obtained by ground-based observatories and spacecraft and by running it faster than real time on a high-performance computer, this model would be able to make 4- to 7-day predictions of the near-Earth space environment. Utilizing real-time inputs from upstream spacecraft, it should be able to predict many components of the near-Earth space environment several days to several hours beforehand.

4. Dynamic radiation-belt models (i.e., models that respond over time to changing input conditions) are technically feasible and will aid the monitoring and short-term forecasting of conditions near manned and unmanned spacecraft that can be endangered by high particle fluxes. As mentioned, the ability to run a global magnetospheric model at a forecast center is anticipated within the next couple of years. Once observational tests of the radiation-belt model (with input from the global MHD simulation) show that it has sufficient accuracy, the model should be put into real-time operation and made available for use in the ISS program. Sufficient resources should therefore be devoted to developing a dynamic theoretical model of the evolution of the radiation belts in time-dependent magnetospheric electromagnetic fields calculated by a global three-dimensional MHD model of the magnetosphere driven by measurements of the solar wind.

5. If observational tests show merit, a scheme should be developed and implemented for calculating a real-time, data-driven map of the radiation belts that uses as input observations by available monitoring spacecraft.

A.6.2 Projects Deserving Early Attention

The following projects, which should eventually contribute to reducing radiation risk in ISS operations, merit early attention from one or more of the affiliated agencies of the National Space Weather Program:

1. Quantitative, theoretical attacks on the particle acceleration problem are sorely needed. Numerical simulations alone can illuminate the details of the acceleration process under realistic conditions, but such efforts have only just begun. Aspects of shock acceleration in SPEs still elude quantitative understanding. Among the less mature areas of understanding in the field of Sun-Earth connections, this area is particularly critical. Until we understand how particles are accelerated at shocks, there can be no first-principles numerical code for predicting SPE parameters from solar inputs.

2. Projects aimed at understanding the underlying physics of CMEs and solar flares are needed for developing an end-to-end capability for forecasting the hazards of radiation for humans in space.

3. It is time to move beyond existing one- and two-dimensional models to three-dimensional MHD simulations in order to correctly model the initiation of solar storms (and the associated shocks), their propagation through the interplanetary medium, and their impact on the near-Earth environment. Observational inputs, preferably real-time, to initialize and drive MHD simulations are now feasible and can impart much-needed realism to the modeling results. The combination of heliospheric modeling with real-time solar data could be used to predict arrival times and magnetic-field characteristics of CMEs directed toward Earth with a 1- to 2-day lead time.

4. Most of the impulsive SPEs reach Earth nearly as rapidly as any electromagnetic signatures (~8 to 80 minutes), leaving insufficient time to make appropriate changes in most EVAs. As a result, flare monitoring with present and upcoming spacecraft will provide only short-term warning, at best, of oncoming energetic particles from such events. To lengthen the lead time available for incorporating information on flare occurrence into EVA scheduling, flare prediction models should be given increased attention.

A.7 NOTES AND REFERENCES

1. *The National Space Weather Program: The Strategic Plan, August 1995,* Office of the Federal Coordinator for Meteorological Services and Supporting Research, Silver Spring, Md.

2. M. Dryer, "Coronal transient phenomena," *Space Sci. Rev.,* 33, 1982, p. 233.

3. N.U. Crooker, J.A. Joselyn, and J. Feynman, eds., *Coronal Mass Ejections, Geophys. Monogr. Ser.,* 99, American Geophysical Union, Washington, D.C., 1997. See especially the chapters by Low; Mikic and Linker; Chen; and Wu and Guo.

4. P. Sturrock, "The role of eruption in solar flares," *Solar Phys.,* 121, 1989, p. 387.

5. U.S. Antiochos, "The magnetic topology of solar eruptions," *Astrophys. J.,* L181, 1998, p. 502.

6. M. Dryer, "Interplanetary studies: propagation of disturbances between the Sun and the magnetosphere," *Space Sci. Rev.,* 67, 1994, p. 363.

7. D. Ostrcil and V. Pizzo, "Three dimensional propagation of CMEs in a structured solar wind flow, 1. and 2.," *J. Geophys. Res.,* 104, 1998, pp. 483 and 493.

8. H.V. Cane, "The current status of our understanding of energetic particles, mass ejections, and flares," in *Coronal Mass Ejections,* Crooker, Joselyn, and Feynman, eds., 1997, pp. 205-215.

9. See articles by D. Reames and M. Lee in *Coronal Mass Ejections,* Crooker, Joselyn, and Feynman, eds., 1997. A more recent reference is D.V. Reames, "Particle acceleration at the sun and in the heliosphere," *Space Sci. Rev.,* in press.

10. D.M. Sawyer and J.I. Vette, "AP8 trapped proton environment for solar maximum and solar minimum," *NSSCE 76-06,* NASA Goddard Space Flight Center, 1976.

11. J.I. Vette, "AE8 trapped electron environment," *NSSDC91-24,* NASA Goddard Space Flight Center, 1991.

12. D.H. Brautigam, M.S. Gussenhoven, and E.G. Mullen, "Quasi-static model of outer zone electrons," *IEEE Trans. Nucl. Sci.,* 39, 1992, p.1797.

13. M.S. Gussenhoven, E.G. Mullen, and D.H. Brautigam, "Improved understanding of the earth's radiation belts from the CRRES satellite," *IEEE Trans. Nucl. Sci.,* 43, 1996, p. 353.

14. K.J. Kerns and M.S. Gussenhoven, CRRESRAD Documentation, PL-TR-92-2201, Phillips Laboratory, AFMC, Hanscomb Air Force Base, 1992.

15. M.S. Gussenhoven, E.G. Mullen, J.T. Bell, D. Madden, and E. Holeman, "APEXRAD: Low altitude orbit dose as a function of inclination, magnetic activity and solar cycle," *IEEE Trans. Nucl. Sci.,* 44, 1997, p. 2161.

16. D.N. Baker, T.I. Pulkkinen, X. Li, S.G. Kanekal, J.B. Blake, R.S. Selesnick, M.G. Henderson, G.D. Reeves, H.E. Spence, and G. Rostoker, "Coronal mass ejections, magnetic clouds, and relativistic magnetospheric electron events: ISTP," *J. Geophys. Res.*, 103, 1998, pp. 17279-17291.

17. M.K. Hudson, S.R. Elkington, J.G. Lyon, V.A. Marchenko, I. Roth, M. Temerin, J.B. Blake, M.S. Gussenhoven, and J.R. Wygant, "Simulations of radiation belt formation during storm sudden commencements," *J. Geophys. Res.*, 102, 1997, pp. 14087-14102.

18. G.D. Reeves, R. Friedel, and R. Hayes, "Maps could provide space weather forecasts for the inner magnetosphere," *EOS*, 79, 1998, p. 613.

19. For a general introduction to the uses of neural networks in applied space weather, see *Proceedings of the International Workshop on Artificial Intelligence Applications in Solar-Terrestrial Physics*, Lund, Sweden, September 22-24, 1993, J. Joselyn, H. Lundstedt, and J. Trolinger, eds., available from NOAA Space Environment Laboratory, Boulder, Colo.

20. T. Aso and T. Ogawa, "Introduction of neural network in the short-term prediction of solar flares," in *Proceedings of the International Workshop on Artificial Intelligence Applications in Solar-Terrestrial Physics*, J.A. Joselyn, H. Lundstedt, and J. Trolinger, eds., available from NOAA Space Environment Center, Boulder, Colo., 1993, pp. 77-82.

21. G.M. Forde, L.W. Townsend, and J.W. Hines, "Application of artificial neural networks in predicting astronaut doses from large solar particle events in space," in *Proceedings of the ANS Topical Conference on Technologies for the New Century, Vol. I*, Nashville, Tenn., April 19-23, 1998, pp. 523-529.

22. See two reviews by A.S. Sharma: "Assessing the magnetosphere's nonlinear behavior: Its dimension is low, its predictability, high," *Reviews of Geophysics, Supplement*, July 1995, pp. 645-650; and "Nonlinear dynamical studies of global magnetospheric dynamics," in *Nonlinear Waves and Chaos in Space Plasmas*, T. Hada and H. Matsumoto, eds., Terra Scientific Publishing Company, Tokyo, 1997, pp. 359-389.

23. G.A. Stringer and R.L. McPherron, "Neural networks and predictions of day-ahead relativistic electrons at geosynchronous orbit," in *Proceedings of the International Workshop on Artificial Intelligence Applications in Solar-Terrestrial Physics*, J.A. Joselyn, H. Lundstedt, and J. Trolinger, eds., available from NOAA Space Environment Center, Boulder, Colo., 1993, pp. 139-143.

24. R.A. Howard, D.J. Michels, N.R. Sheeley, Jr., and M.J. Koomen, "The observation of a coronal transient directed toward the Earth," *Astrophys. J.*, 1982, p. L101.

25. J.A. Joselyn, "Geomagnetic activity forecasting: The state of the art," *Rev. Geophys.*, 33, 1995, p. 383.

26. See note 8.

27. See note 1.

28. See note 1.

29. See note 1.

30. N.A. Tsyganenko, "A magnetospheric magnetic field model with a warped tail current sheet," *Planet. Space Sci.*, 37, 1989, pp. 5-20.

APPENDIX B

Statement of Task

Background: NASA's Sun-Earth Connection (SEC) program office has asked the National Research Council to assess potential impact that energetic particle radiation caused by solar storms might have on scheduling EVAs during the construction phase of the International Space Station (ISS). This assessment bears upon determining the priority the SEC program office might assign to predicting and specifying the intensities of solar energetic particle events and the radiation belts.

The concern that solar storms might non-negligibly affect the scheduling of EVAs arises because NASA plans to place ISS in a circular orbit similar to the Russian MIR at an altitude of approximately 444 km (240 nmi) with a 51.6 degree inclination to the equatorial plane. This inclination is higher than planned prior to the decision to collaborate with Russia. It places ISS in a space environment that is vulnerable to space weather disturbances from auroral influences on the upper atmosphere, from solar and interplanetary energetic particles, and from the galactic cosmic ray flux. Concern about the potential levels of radiation exposure by Station workers and crew, especially those engaged in work outside Station modules, is heightened by a timetable for construction that is scheduled to occur during the upcoming period of maximum of solar activity. (Scientific considerations for the next solar activity maximum are addressed in the new NRC report "Readiness for the Upcoming Solar Maximum," in press.) The higher inclination orbit also increases vulnerability to "single event upsets" in solid-state devices and electrical discharges on material surfaces in space when the sun is active.

NASA plans to rely on NOAA's Space Environment Center (NOAA/SEC) in Boulder, Colorado, for space weather warnings. However, NOAA/SEC does not provide services that are tailored specifically for Space Station needs. Moreover, there is no group within NASA that is specifically supported to "translate" the more general NOAA/SEC space weather products for ISS operations use.

Plan: A study will be undertaken that will examine, on the one hand, the risk of radiation exposure to ISS and, on the other hand, NASA's plans to use SEC space weather information to manage the radiation risk. The Committee on Solar and Space Physics, working in conjunction with its federated partner, the Committee on Solar-Terrestrial Research, will take the lead in producing a report. The study will make recommendations that take advantage of current research in the Solar Connections Enterprise in OSS and the developing interagency National Space Weather Program. A report of the order of 25 pages is planned.

It is anticipated that the report would accomplish the following objectives:

Provide a preliminary assessment of the radiation exposure impact to ISS assembly activities. This aspect of the study will be executed in two parts. In the first part, the committees will simulate the radiation exposure to ISS during the coming solar maximum (1998-2002) by shifting the radiation events of the last solar cycle (cycle 22) ahead one cycle. This simulation is justified since the consensus prediction for solar cycle 23 (during which ISS will be constructed) is that it will be very similar to solar cycle 22 (for which there are data on radiation intensities from solar storms). The radiation events will be superimposed on the Space Shuttle flight and EVA (extravehicular activities) schedule for ISS assembly. For the second part of the study, the committees will estimate how often "significant" disruption to activities might occur. This first-order estimate will not be based on a detailed analysis of the radiation dose to an astronaut. Instead, the committees will first establish thresholds (based on intensity and duration) which, if exceeded, would be expected to disrupt the mission schedule. Then they will examine solar events from cycle 22 to determine how often this threshold was exceeded during scheduled EVAs. This exercise will be repeated several times with different start times for the assembly flights to obtain statistics with which to express the impact assessment.

Examine existing arrangements within NASA to manage the radiation risk problem. This fact-finding part of the study would provide background information on ongoing and planned activities at NASA centers and at headquarters. It would also examine arrangements between the NASA science and life sciences codes.

Recommend how operational radiation support to ISS might be optimally accomplished. This part of the study would focus on current capabilities and planned roles and responsibilities in NASA, NOAA, and DOD for generating, disseminating, and using operational space weather products, and it would identify gaps or overlaps and opportunities for improving support.

Meetings: The Committee on Solar and Space Physics, working in conjunction with its federated partner, the Committee on Solar-Terrestrial Research, will take the lead in this study. The committees have considerable expertise related to space weather and its potential effects on Space Station. The study is planned as a 1-year effort with three 3-day meetings for further fact finding, discussions, and drafting the report. The first meeting will be held on June 29-July 1, 1998. Subsequent meetings would be held in the fall of 1998 and late January/early Febuary 1999. The final meeting to complete the report would likely be held at the Beckman Center. Publication of the report would occur by July 1, 1999.

APPENDIX C

Biographies of Committee Members

COMMITTEE ON SOLAR AND SPACE PHYSICS

GEORGE L. SISCOE, Chair, has been a senior research professor at Boston University's Center for Space Physics since 1993. His research interests center around the large-scale organization and dynamics of plasmas and fields in space-solar wind, planetary magnetospheres, and, in particular, Earth's magnetosphere. He is active in research in support of the National Space Weather Program. Earlier, he was on the teaching faculty of the Department of Atmospheric Sciences for 25 years, during 7 of which he served as chair of the department. He was formerly a member of the Space Studies Board, the Board's Committee on Planetary and Lunar Exploration, and chair of NASA's Space Physics Advisory Committee. He is currently a member of the UCAR Advisory Panel for the National Centers for Environmental Predictions and a fellow of the American Geophysical Union.

CHARLES W. CARLSON is a research physicist and senior space fellow at the Space Sciences Laboratory, University of California, Berkeley. Dr. Carlson has over 30 years experience in magnetospheric and space plasma physics research with over 85 publications. He has also developed many of the plasma instruments currently used in this area of research. These experiments include numerous sounding rockets to study plasma in the auroral zone and equatorial ionosphere as well as the Giotto mission to comet Halley, AMPTE, and Mars Observer. He is currently the principal investigator for the Fast Auroral Snapshot (FAST) satellite, a NASA small-class explorer (SMEX) mission launched in August 1996 that is providing data on the processes thought to be responsible for producing Earth's aurora. Dr. Carlson is also a coinvestigator on the Wind and POLAR missions.

ROBERT L. CAROVILLANO is a member of the Boston College faculty. In space physics research, Dr. Carovillano has published on a broad spectrum of topics in pure theory and data analysis, including magnetospheric energy theorems and related topics. Dr. Carovillano has served on national advisory committees of the National Academy of Sciences, the National Center for Atmospheric Research, the National Aeronautics and Space Administration (NASA), and the National Science Foundation (NSF) and has chaired several such advisory committees. Dr. Carovillano has been principal investigator on many research grants and contracts funded by the NSF, NASA, the Office of Naval Research, and the U.S. Air Force. Since July 1994, Dr. Carovillano has been a visiting senior scientist at NASA headquarters in the Office of Space Science. At NASA he has been responsible for the supervision of several programs and research initiatives in space physics but has been most deeply engaged in optimizing mission scientific accomplishments and opportunities.

TAMAS I. GOMBOSI is presently senior editor of the *Journal of Geophysical Research—Space Physics*. In the mid 1970s he was the first foreign national to do postdoctoral research at the Space Research Institute in Moscow, where he participated in the data interpretation of the Venera-9 and Venera-10 Venus orbiters. A few years later he came to the United States to participate in theoretical work related to NASA's Venus exploration. In the early 1980s he played a leading role in the planning and implementation of the international VEGA mission to Venus and Halley's comet. He played a pioneering role in the development of modern cometary plasma physics. Also, he was among the first scientists to explain the acceleration of pickup ions by self-generated, low-frequency MHD waves. Dr. Gombosi is author or coauthor of about 150 peer-reviewed scientific publications and over 250 presentations.

RAYMOND A. GREENWALD is supervisor of the Ionospheric and Atmospheric Remote Sensing Group in the Space Sciences Branch of the Applied Physics Laboratory at Johns Hopkins University. He is active in the ISTP community and is widely known for the design and development of the STARE radar system in northern Scandinavia, which studies the circulation of the high-latitude ionosphere. He is also associated with the development of the extensive high-latitude network of HF radars known as SuperDARN and is currently chairman of the international SuperDARN Executive Committee.

JUDITH T. KARPEN is a research astrophysicist in the Solar-Terrestrial Relationships Branch of the Space Science Division of the Naval Research Laboratory. Her primary research interests include analytical and numerical modeling of dynamic solar and heliospheric phenomena and applications of plasma physics and magnetohydrodynamics to solar and astrophysical activity. She has been involved in analysis and interpretation of solar data obtained with the hard X-ray burst spectrometer on OSO-5 and with the NRL X-ray spectrometer (SOLPLEX) and white-light coronagraph (SOLWIND) on board the P78-1 satellite. Since 1984, Dr. Karpen has been a coinvestigator or principal investigator in numerous research projects sponsored by NASA, AFGL, and DOD. She is currently chair of the AAS Solar Physics Division and was a member of the MOWG advising the NASA Solar Physics Branch for several years.

GLENN M. MASON is professor jointly in the Department of Physics and the Institute for Physical Science and Technology at the University of Maryland, College Park. He has worked on the development of novel instrumentation that allows determination of the mass composition of solar and interplanetary particles in previously unexplored energy ranges. His research work has included galactic cosmic rays, solar energetic particles, and the acceleration and transport of particles both in the solar atmosphere and in the interplanetary medium. He is principal investigator on the NASA Solar, Anomalous, and Magnetospheric Explorer (SAMPEX) spacecraft mission and is coinvestigator on energetic particle instruments for the NASA Wind spacecraft and the NASA Advanced Composition Explorer (ACE) spacecraft. He was formerly chair of the NASA Sun-Earth Connections Advisory Subcommittee (SECAS) and the NASA Space Science Advisory Committee (SScAC) and is currently a member of the NAS/NRC Committee on Solar and Space Physics.

MARGARET A. SHEA has worked in the geophysics division of the Air Force laboratory at Hanscomb Air Force Base since 1964. She has received numerous Air Force awards for superior performance and in 1985 was a recipient of the Air Force Geophysics Laboratory Guenter Loeser Memorial Award for outstanding career contributions. She was elected to be an associate of the Royal Astronomical Society (equivalent to fellow) in 1991, and in 1995 she was elected as corresponding member of the International Academy of Astronautics.

KEITH T. STRONG is currently the manager of the Lockheed Martin Solar and Astrophysics Laboratory and director of the Space Science Independent Research Program. Also, since 1991 and 1994, respectively, he has been coinvestigator on the Yohkoh Soft X-ray Telescope and deputy principal investigator (science) on the Transition Region & Coronal Explorer. He specializes in the use of diagnostics derived from X-ray lines to characterize the temperature, emission measure, abundances, and dynamics of coronal plasmas to improve our understanding of active regions and how they evolve and produce flares. Dr. Strong served on the Sun-Earth Connection Roadmap

Planning Committee (NASA), 1996-1997; the Solar-B Science Definition Team (NASA), 1996-1997; the STEREO Science Definition Team (NASA), 1997; and has been a member of the NRC-CSSP since 1997. He has over 200 publications in refereed journals and over 200 talks and public lectures.

RICHARD A. WOLF joined the Rice faculty in 1967, after completing his postdoctoral work at Caltech and serving a year on the technical staff of Bell Laboratories. Although his early research was in theoretical astrophysics, he now works primarily on the plasma physics of the solar system, concentrating on the space near the Earth. He is best known for his work with the Rice Convection Model, which is a large computer code representing plasma motions in Earth's magnetosphere and ionosphere. For the past 7 years, Dr. Wolf has participated in the development of the Magnetospheric Specification Model that will soon be placed in service at the USAF Space Forecast Center, providing data for reports on space weather. His research group is participating in two national efforts to develop comprehensive research computer models of Earth's magnetosphere. Currently, Dr. Wolf is finishing a book on Earth's magnetosphere.

COMMITTEE ON SOLAR-TERRESTRIAL RESEARCH

MICHAEL C. KELLEY, Chair, is professor of electrical engineering at Cornell University (1975-present). He received his Ph.D. in electrical engineering from the University of California, Berkeley, in 1970. Half of his research effort involves using radar-lidar observatories to measure wind and wave patterns from 30 to several hundred kilometers above the surface of Earth. The rest involves the use of satellites and rockets to carry Cornell instrumentation directly into the space environment to study in detail specific atmospheric phenomena such as thunderstorms and the aurora. Dr. Kelley is a fellow of the American Geophysical Union and in 1979 won that society's James B. Macelwane Award.

MAURA E. HAGAN is a scientist at the High Altitude Observatory of the National Center for Atmospheric Research (1992-present). She received her Ph.D. in physics from Boston College in 1987. Her research interests include the physics of the upper atmosphere; chemical/dynamical coupling between the mesosphere, lower thermosphere, upper thermosphere, and ionosphere; atmospheric tides and waves; electrodynamic coupling between ionospheric and magnetospheric plasma and the neutral thermosphere; global change as it pertains to the upper atmosphere; numerical modeling of the upper atmosphere system; and analysis and interpretation of incoherent scatter radar measurements. Dr. Hagan is a member of the American Geophysical Union and the American Association for the Advancement of Science.

MARY K. HUDSON is a professor in and chair of the Department of Physics and Astronomy at Dartmouth College. Her professional experience includes work as a research physicist at the Aerospace Corporation from 1969 to 1971 and as a research physicist in space physics at the University of California, Berkeley. Dr. Hudson's research interests include theoretical models of ionosphere plasma phenomena; E and F region irregularities; ionosphere-magnetosphere coupling and transport phenomena; and ring current-plasma pause interaction and other planetary magnetospheres. She is a recipient of the Macelwane Award from the American Geophysical Union.

NORMAN F. NESS is president and professor of physics at the Bartol Research Institute, University of Delaware, Newark (1987-present). He received his Ph.D. in geophysics from the Massachusetts Institute of Technology in 1959. His specialty is space physics. His research emphasizes experimental investigation of magnetic fields. Dr. Ness is a fellow of the American Geophysical Union. He has received the Arthur S. Flemming Medal for the U.S. Government (1968); the Space Science Award, American Institute of Aeronautics and Astronautics (1972); and the John Adam Fleming Medal of the American Geophysical Union (1965).

THOMAS F. TASCIONE is vice president, Space and Environmental Systems Operations, Sterling Software (1966-present). In this position he is leading an effort to start a commercial space weather forecasting center. He received his Ph.D. in space physics from Rice University (1982). His prior professional experience was with the Department of Defense (1972-1993). As deputy director of the Air Force Weather Service, he served as the focal point on all space weather activities. He co-chaired the interagency committee that initiated and developed the National Space Weather Program (NSWP). He was instrumental in the development of the NSWP Strategic and Implementation Plans. During his Air Force career, Dr. Tascione served as the lead space weather forecaster and architect of the Air Force space weather forecast models program.

APPENDIX D

Acronyms and Abbreviations

ACE	Advanced Composition Explorer (NASA research spacecraft)
AFRL	Air Force Research Laboratory
ALARA	as low as reasonably achievable (operational flight rule for radiation management)
APEXRAD	model of radiation dose based on APEX satellite data
BFO	blood-forming organs
CEPPAD	comprehensive energetic particle and pitch angle distribution
CIR	corotating interaction regions
CME	coronal mass ejection
CPD	crew passive dosimeter
CRRES	Combined Release and Radiation Effects Satellite (NASA research satellite)
CRRESELE	model of electron radiation belts based on CRESS data
CRRESPRO	model of proton radiation belts based on CRRES data
CRRESRAD	model of radiation doses based on CRRES data
DMSP	Defense Meteorology Satellite Program (a suite of polar-orbiting DOD satellites)
DOD	Department of Defense
DOE	Department of Energy
DOI	Department of the Interior
EIT	Extreme Ultraviolet Imaging Telescope
ESA	European Space Agency
EUV	extreme ultraviolet
EVA	extravehicular activity—astronauts in space suits outside a shuttle or station
FAST	Fast Auroral Snapshot Explorer

GCR	galactic cosmic ray
GeV	gigaelectronvolt (unit of energy = 1,000 MeV)
GOES	Geostationary Operational Environmental Satellites (NOAA)
GSFC	Goddard Space Flight Center
Gy	gray (unit of radiation)
HESSI	High Energy Spectroscopic Imager (NASA research spacecraft)
HRE	highly relativistic electron
HZE	high z energetic particle, i.e., high-atomic-number particle
IMEX	Inner Magnetospheric Explorer
IMF	interplanetary magnetic field
IMP	Interplanetary Monitoring Program
ISS	International Space Station
ISTP	International Solar-Terrestrial Physics (a NASA program)
JPL	Jet Propulsion Laboratory
JSC	Johnson Space Center (the NASA center responsible for ISS and shuttle operations)
Kp	index of geomagnetic activity
L1	First Lagrangian point. Lagrangian points are points between two orbiting masses in which the gravitational pulls from both bodies are balanced exactly with the centripetal force required to rotate with them. Objects at these points then orbit at a constant distance from both masses.
LASCO	Large Angle and Spectrometic Coronagraph Experiment
LEM	lunar excursion module
LEO	low earth orbit, referring to a class of satellites
MCC	Mission Control Center
MeV	million electron volts (unit of energy)
MHD	magnetohydrodynamic
MSC	Manned Spacecraft Center
MSFC	Marshall Space Flight Center
NASA	National Aeronautics and Space Administration
NCRP	National Council on Radiation Protection and Measurements
NOAA	National Oceanic and Atmospheric Administration
NRA	NASA Research Announcement
NSF	National Science Foundation
NSSDC	National Space Science Data Center, located at GSFC
NSWP	National Space Weather Program
OTTI	Orbiting Technology Testbed Initiative (program within SEE)
POES	Polar-Orbiting Operational Environmental Satellites (NOAA)
POLAR	a high-altitude, polar-orbiting research satellite (NASA)
RBE	relative biological effectiveness
RPC	Rapid Prototyping Center at SEC and AFRL for expediting the transitioning of research into operations

SAA   South Atlantic Anomaly
SAMPEX   Solar Anomalous Magnetospheric Particle Explorer (NASA research satellite)
SEC   Space Environment Center (NOAA)
SEE   Space Environment Effects program at MSFC
SOHO   Solar and Heliospheric Observatory (ESA/NASA research satellite)
SPAN   Solar Particle Alert Network
SPE   solar particle event
SRAG   Space Radiation Analysis Group (at Johnson Space Center)
Sv   sievert (unit of radiation)
SWOP   Space Weather Operations Center (at NOAA's SEC)
STEREO   Solar Terrestrial Relations Observatory (NASA research spacecraft)
SXI   Solar X-ray Imager

TCEP   tissue-equivalent proportional counter

USAF   United States Air Force